The TransForum Model: Transforming Agro Innovation Toward Sustainable Development

Henk C. van Latesteijn • Karin Andeweg
Editors

The TransForum Model: Transforming Agro Innovation Toward Sustainable Development

Springer

Editors
Henk C. van Latesteijn
TransForum
P.O. Box 80
2700 AB Zoetermeer
The Netherlands
VanLatesteijn@TransForum.nl
and
Louis Pasteurlaan 6
2719 EE Zoetermeer
The Netherlands

Karin Andeweg
TransForum
Louis Pasteurlaan 6
2700 AB Zoetermeer
The Netherlands
Andeweg@TransForum.nl

ISBN 978-94-007-9011-7 ISBN 978-90-481-9781-1 (eBook)
DOI 10.1007/978-90-481-9781-1
Springer Dordrecht Heidelberg London New York

Softcover re-print of the Hardcover 1st edition 2011

Front Cover Picture: "Metropolitan Agriculture" Wall-Painting by Thomas Moes
Cover design: deblik

Printed on acid-free paper

Springer is part of Springer Science+Business Media (www.springer.com)

Table of Contents

Preface ... **xi**

Chapter 1. The need for a new agro innovation system ... **1**

1.1. Introduction: The challenge ... 1
1.2. Innovating in the agricultural system ... 3
1.3. Sustainable development as an innovation target ... 6
1.4. Characteristics of innovation and the innovation system ... 7
1.5. System innovation as a process of engagement ... 10
1.6. Five motivating assumptions for enhancing sustainable development in agriculture ... 13
Assumption 1: Sustainable development is a dynamic process ... 14
Assumption 2: Sustainable development needs system innovation ... 14
Assumption 3: System innovation is a non-linear learning process ... 15
Assumption 4: System innovation requires a multi-stakeholder approach ... 15
Assumption 5: Multi-stakeholder approaches imply trans-disciplinary knowledge creation ... 15
1.7. Stimulating sustainable development in a learning-by-doing experiment ... 16
1.8. References ... 17

Chapter 2. Inventions for future sustainable development in agriculture ... **21**

2.1. Inventions – an introduction ... 21
2.2. Inventions in agricultural infrastructure ... 22
2.2.1. Ownership issues: Inventions, innovations and IPR ... 23
2.2.1.1. Ownership and agricultural inventions of the past ... 23
2.2.1.2. Ownership and modern inventions in agricultural sciences ... 25
2.2.1.3. Open innovation under attack ... 26
2.2.2. Organization of inventions and innovation ... 27
2.2.2.1. 'Ware' elements of inventions and entrepreneurship ... 27
2.3. Three ways to stimulate needed inventions ... 29
2.3.1. Intersectional inventions ... 30
2.3.1.1. Organic agriculture ... 30
2.3.1.2. GM crops ... 31
2.3.2. Directional inventions ... 32
2.3.2.1. Cisgenesis ... 32
2.3.3. Open innovation ... 33
2.3.3.1. Care farming ... 34
2.3.3.2. Energy producing greenhouses ... 34

2.4. Conclusion ... 35
2.5. References ... 37

Chapter 3. Organizing innovations and transitions ... 41

3.1. Introduction: Context of the research ... 41
3.2. Theoretical building blocks ... 43
3.3. Coming in from the space of flows ... 45
3.3.1. Innovation as reconfiguration ... 46
3.3.2. Between 'business as usual' and 'market failures' ... 47
3.4. Where the space of flows meets the space of places ... 50
3.4.1. Perspective from the space of places ... 51
3.4.2. The importance of a 3P regional business case ... 53
3.5. Conclusion: Breaking the developmental lock-in ... 54
3.6. References ... 56

Chapter 4. For or against innovation? The influence of images ... 59

4.1. Introduction ... 59
4.2. Images in innovation ... 60
4.3. Hypotheses and research approach ... 62
4.3.1. Hypotheses ... 62
4.3.2. Research approach ... 63
4.4. The role of images – case studies ... 64
4.4.1. Images within innovation projects – Northern Frisian Woods ... 64
4.4.2. Societal influences on innovation experiments – New Mixed Farm ... 66
4.5. Dealing with images in innovation ... 67
4.5.1. Collaborative visualization ... 67
4.5.2. Managing images ... 69
4.6. Conclusion ... 70
4.7. References ... 71

Chapter 5. Mobilizing consumer demand for sustainable development ... 73

5.1. Introduction ... 73
5.2. Attitude-behavior gaps ... 77
5.2.1. Limitation to the theory of planned behavior ... 77
5.2.2. Socio-temporal dilemmas in choice ... 78
5.2.3. Construal level theory ... 79
5.2.4. Multiple selves ... 80
5.3. Stimulating sustainable consumer behavior ... 81
5.4. Facilitating the three routes of mobilizing sustainable consumption ... 84
5.4.1. Route 1: Facilitating motivation ... 84

5.4.1.1. The multiple selves in sustainable behavior 84
5.4.1.2. Social norms in sustainable development 84
5.4.1.3. Implicit factors in behavior aimed at sustainable development 86
5.4.1.4. The willingness to pay for sustainability labels and logo 87
5.4.2. Route 2: Facilitating ability 87
5.4.3. Route 3: Facilitating opportunity 88
5.4.3.1. Corporate social responsibility activities 88
5.4.3.2. Enhancing opportunity from the primary production level 89
5.4.3.3. Facilitating opportunity and consumer choice at the point of purchase 89
5.4.4. Methodological challenges in stimulating consumer demand 90
5.5. Key lessons learned 91
5.6. Conclusion 93
5.7. References 93

Chapter 6. From motivating assumptions to a practical innovation model ... 97

6.1. Introduction 97
6.1.1. Measuring and explaining performance 98
6.2. Methodology 99
6.3. Performance of cases and assumptions 101
6.3.1. Performance groups 101
6.3.2. Performance of the five motivating assumptions 103
6.4. Influence of explanatory attributes and links to motivating assumptions.. 105
6.4.1. Explanatory attribute categories and performance 106
6.4.2. Explanatory attributes, core strategies and the motivating assumptions 107
6.4.2.1. Influence of initiating conditions 108
6.4.2.2. Influence of attributes related to assumption 4 (multi-stakeholder engagement) 109
6.4.2.3. Influence of attributes related to motivating assumptions 2, 3, and 5 111
6.4.3. Key findings 113
6.5. From motivating assumptions to a practical innovation model for innovation 114
6.5.1. Core strategy I: Action experiment formation and guidance 115
6.5.2. Core strategy II: Process monitoring and reflection 116
6.6. Key learnings from working with the motivating assumptions and core strategies 117
6.6.1. Dealing with the wickedness of sustainability in setting project vision and ambition 117
6.6.2. Dealing with inherent complexity 118

6.6.3. Dealing with knowledge creation in multi-stakeholder groups ... 119
6.6.4. Dealing with monitoring and reflection ... 121
6.6.4.1. Dynamic Learning Agenda ... 122
6.6.4.2. Interactive learning and reflexive monitoring in action ... 122
6.6.5. Dealing with the tension between a project and its institutional context ... 123
6.7. Applying the TransForum model ... 124
6.7.1. A traveler's guide to managing sustainable development ... 124
6.7.2. Implications with regard to the current innovation system ... 125
6.7.3. Broader application of the model ... 126
6.8. References ... 127

Chapter 7. Transforming agriculture: A new approach to stimulate sustainable development. ... 131

7.1. Introduction ... 131
7.2. The TransForum model ... 133
7.2.1. From assumptions to principles ... 133
7.2.2. The sixth principle ... 135
7.2.3. Accompanying strategies ... 137
7.3. Contributions from scientific disciplines ... 139
7.3.1. The Natural Sciences lens ... 139
7.3.2. The Social Sciences lens ... 141
7.4. A promising perspective ... 143
7.5. References ... 144

List of Abbreviations ... 145

List of Contributors ... 147

Appendix I. TransForum: Stimulating innovation for sustainable development by learning by doing ... 149

Action experiments ... 150
Scientific research ... 150
Learning projects ... 150

Appendix II. Action experiments explicated ... 153
Action experiment SynErgy ... 153
Action experiment more about food ... 153
Action experiment regional food chain ... 154
Action experiment Green Care ... 154
Action experiment new markets and vital coalitions Heuvelland ... 154
Action experiment New Mixed Farm ... 155

Action experiment saline agriculture 155
Action experiment Northern Frisian Woods 156

Appendix III. Overview experiments and projects 157

Action experiments 157
Scientific research projects 158
Learning projects 158

Appendix IV 161

Preface

Why does so much research not result in beneficial applications in practice? This question, sometimes referred to as the knowledge paradox, has been raised in the political arena and has resulted in developing and testing new procedures to initiate and fund innovative research on sustainable development. Classic research tends to be mono-disciplinary while literature suggests that societal questions associated with sustainable development may require an interdisciplinary approach including substantial interaction with stakeholders and policy makers. The TransForum innovation program, results of which are reported in this book, was organized in such a new way. Not through regular governmental or university channels or through the Dutch National Science Foundation, but on the basis of a special Governmental Fund filled with money originating from the sale of natural gas to customers outside the Netherlands. These funds were put aside to stimulate innovative research. In 2004, 800 million euros were dedicated to a national research program on innovation. One subgroup was to study sustainable development and TransForum – focusing on the agricultural sector – was one of the programs being funded in that category for 30 million euros, to be matched by societal partners at 50%. The scientific quality of these programs, including the one of TransForum, was judged by committees of the Royal Dutch Academy of Sciences, Arts and Letters and the relevance for society by the Central Planning Agency of the Netherlands. Programs were only approved when both judgments were positive. This was also a new procedure as assessments of research programs are traditionally based on peer review only.

This book is published at the end of the program period in 2010 and reflects activities of the scientific research program of TransForum consisting of 25 research projects. Their identification and set-up also followed an original, unusual approach. The overall TransForum program was characterized by demand-orientation: problems to be studied were identified and submitted to TransForum by entrepreneurs in agribusiness and various stakeholder groups. Overall, 33 so-called action experiments were identified and much time was spent in discussing and defining possible research needs when solving problems related to the experiments. Such discussions are not easy, require new abilities that go beyond those associated with classic research and take a lot of time. Understandably, some researchers are therefore less than enthusiastic to embrace this new interactive approach which is particularly relevant when studying so-called 'wicked' problems involving many stakeholders with widely differing goals and viewpoints, while simple solutions do not exist.

The following chapters present, therefore, the results of a continuing struggle and certainly do not offer any magic clues as to perfect procedures for future research on 'wicked' problems. But results should be helpful for scientists and others struggling to develop effective research procedures that are both scientifically robust, while offering essential input in solving societal problems of the 21th century. Here emphasis is on agricultural development but results obtained are also relevant for research in general.

Prof. Dr. Johan Bouma
Em. Professor of Soil Science, Wageningen University
Chair Advisory Board TransForum.

Chapter 1. The need for a new agro innovation system

Henk C. van Latesteijn[1] and Karin Andeweg[2]

[1]VanLatesteijn@TransForum.nl; TransForum, P.O. Box 80, 2700 AB, Zoetermeer, The Netherlands

[2]Andeweg@TransForum.nl; TransForum, P.O. Box 80, 2700 AB, Zoetermeer, The Netherlands

Abstract Agriculture has seen a tremendous development in both production and productivity growth during the last decades. A continuous innovation effort was at the heart of these positive developments. And with that growth, the negative effects on the physical and social environment have also become apparent. For the coming decades the world population will continue to grow, and rising incomes will lead to the increasing demand for food and other agricultural produce. These developments urge the agricultural sector to shift toward a more sustainable development so that production will continue to rise and the impact on the environment will diminish. Systems innovation is needed to bring this about. Based on a large number of action experiments we propose five assumptions that may alter the way in which innovation can be organized and stimulated. This book explores the validity and applicability of these assumptions building on an extensive body of research of various disciplines, that was carried out alongside the action experiments. The results are presented as a practical model to transform agro innovation.

1.1. Introduction: The challenge

Innovation in agriculture considerably improved the livelihood of all of us. In the 1950s the ravaged countries after the Second World War needed to rebuild their agricultural production system. At the same time the demand for food was exploding. Through enormous increases in productivity and yields it was possible to feed more people. And even better: innovations in agricultural production and processing made it possible to provide all this food at a decreasing price level. This development was one of the main economic drivers that boosted worldwide economic development by providing an income surplus that could be spent on all sorts of new consumer goods. Never before, the average amount of food per capita was that high (see Figure 1.1).[1]

[1] Please note this concerns the *average* amount of food per capita. This does thus not imply that every person has accessibility to or availability of sufficient food.

H.C. van Latesteijn and K. Andeweg (eds.), *The TransForum Model: Transforming Agro Innovation Toward Sustainable Development*, DOI 10.1007/978-90-481-9781-1_1,

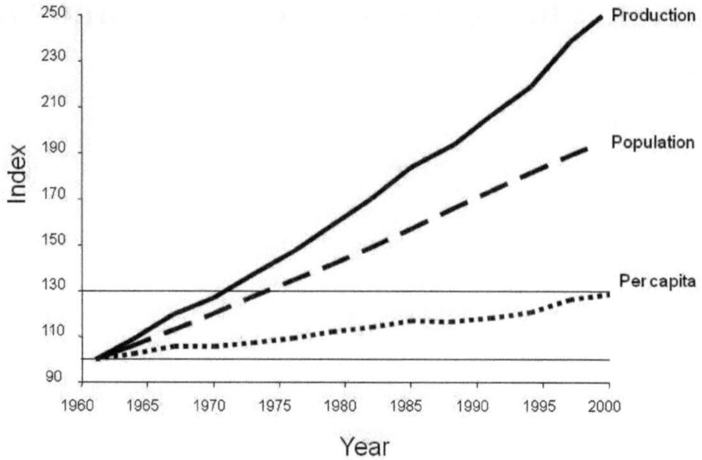

Fig. 1.1. World food production 1960–2000; index 1960=100 (FAO 2002).

However, this development has its dark side also. Increased agricultural production created negative ecological and social side effects (see o.a. Bekman 1998; Jongbloed and Lenis 1998; Schiere and Van Keulen 1999; Wallinga 2009; Boogaard et al. 2010). Increasing yields and an ever rising productivity led to an overuse of non-priced production factors. Water and air quality, as well as soil fertility and natural biodiversity all suffer from overuse and overexploitation. Abundant use of nutrients and pesticides in intensive farming methods and emission of greenhouse gasses is causing environmental pollution, locally and globally. Government, consumers, and citizens have become more critical on different aspects of agricultural developments. In the eyes of a large number of consumers and citizens, the underlying values related to respectfully treating animals, safeguarding the planet and continuing profits are threatened by the way in which agricultural production and processing is executed these days. So, the relationship of agriculture with society has shifted. Ecological and societal limits to agriculture and the larger food production system have been reached (Dunn 2003). This constitutes the core of the sustainability discourse that is facing the agricultural sector with an enormous challenge.

And while these problems are mounting, at the same time the demand for agricultural produce is growing at a very fast rate. In his opening statement for the 2009 World Food Summit Dr. Jaques Diouf, Director-General of the FAO stated that:

> food production must expand by 70 percent in the world and double in developing countries, to meet the food needs of a world population expected to reach 9.1 billion in 2050.

The sustainability debate adds to that challenge the notion that this increase in production must also lead to a substantial increase in societal and environmental quality. This formidable double challenge of expanding production and at the same time improving the way in which this production takes place cannot be tackled by a further intensification of the current system. After all, it is the current system that led us into much of the problems, so with this same system these problems will not be solved. Innovation can help to overcome the negative sides of this development. But this time the innovation must be deeper and more profound, so not only do we need to produce more for an ever growing world population, we also need to produce in a better way, so the negative effects of the past production system will disappear.

Therefore we need a special type of innovation – systems innovation – that transcends the normal operational impact of innovations that focus on improving efficiency. We need innovation of products, processes and contexts that together will lead to 'doing better things', whereas innovation in the business-as-usual context only leads to 'doing things better'. This calls for new mindsets that can come up with new ways of doing things. Through such an approach it will be possible to dramatically change the mode of agricultural production and devise new ways of food production that do not waste renewable resources and that do not put a heavy burden on social and environmental systems.

To sum, agriculture needs to reinvent itself and become a star player again. This can be achieved if we take up the strong innovative forces that exist within the agricultural community and put them to work in another direction. This challenge can be addressed if we transform agro innovation and shift agriculture to a motor of sustainable development.

The question is: How can we do that? What changes are needed? And what changes are absolutely necessary? This introductory chapter further explicates this challenge. We will first expound the characteristics of the agricultural system and the current innovation system, and on the way how sustainable development can be the target of agricultural innovation.

1.2. Innovating in the agricultural system

Agriculture has seen a very rapid growth over the last decades. Triggered by what became known as *the green revolution*, productivity growth of both animal and plant production for human consumption showed a steep increase in the last decades. The success of the green revolution is contributed to a combined effort of research, extension services and education to introduce new varieties, new production and tillage techniques and new modes of cooperation throughout the sector. This linear model is characterized by the close collaboration and even integration of research, education and extension. Inventions in fundamental research in academia form the starting point of a linear trickle-down process through strategic and applied research to knowledge transfer for implementation. Extension services and education to farmers forms the final link in the innovation chain. The result of all these activities becomes visible in the application in the agricultural production and processing system (Leeuwis et al. 2006).

Driving this system was an enormous research effort into more productive and efficient crops and production techniques. Figure 1.2 illustrates the considerable rise in productivity for the most important grain. At national and international levels large investments were made in the knowledge infrastructure to continually improve yield of the different agricultural crops. The CGIAR (Consultative Group for International Agricultural Research) set up commodity based research centers around the globe. Due to the productivity increases brought about by the knowledge generated in this linear research, education and extension system, the world has never had as much food per capita as it has today (see again Figure 1.1). Through these centers research money from richer donor states was transferred to improving basic understanding and knowledge about the world's most important staple food crops for all countries. This is a specific trait of agriculture: we deal here with our daily food and as such we are willing to invest collectively in improving the system as a whole.

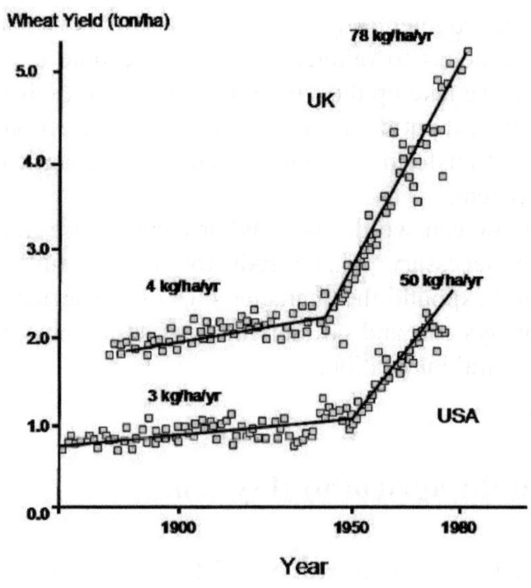

Fig. 1.2. Annual productivity rise of wheat production in the UK and USA, 1850–1980 (Van Latesteijn 1998).

From this perspective, agriculture is not completely comparable to other sectors of the economy. No nation or state can ignore the importance of food production and therefore will be involved in the continuous improvement of the performance of the sector. This specific quality of the agricultural sector provides an opportunity to face the challenge of the need for a more sustainable development. We can tap into the same energy and arguments that gave rise to the green (productivity) revolution and this time use it for a broader purpose. We need to realize that the challenges that are facing us now will not be taken up by acting the same way we have done for decades. This will only improve the efficiency of 'doing the wrong things better'. Instead, we need to find out ways to 'do better things'. This will not be an easy task, since our innovative powers and surrounding institutions are all honed towards the improvement of operational efficiency (Porter 1998). And this operational efficiency is measured in one dimension only: profit.

For sustainable development to occur, we also need more productivity, but it has to be created across more dimensions than just profit. The environmental and social problems have made it clear that we need productivity gains in those areas also. We need more production of valuable agricultural products and services with less impact on environmental and social values. This can only be achieved if we do not incrementally improve what we are already doing, we need transformational change that leads to modes of agricultural production that are surprisingly new. For that we at least need different types of science from a range of disciplines that can provide new answers to challenges related to water use efficiency, robust and pest resistant crops, reduction of post harvest losses and sustainable ways of livestock production (Beddington 2010).

But we also need new arrangements in society itself to accommodate new ideas, arrangements and institutions. Together this can be denoted as a need for 'systems innovation'. In his description of innovation, Porter (1998) makes a clear distinction between invention and innovation. According to Porter innovation is a new way of doing things (so, an invention) that is commercialized. In this sense, innovation is all about 'turning ideas into cash'. For the type of innovation needed to bring about a more sustainable development, the new ideas should not only lead to 'cash', but also to 'environment' and 'social quality'. This needs an extra step that transforms a standard innovation into something more. For that to occur, we need new arrangements in society itself to accommodate new modes of agricultural production not only to lead to new industry and commerce, but also to new relationships with the surrounding physical and social environment.

Innovation as it is commonly understood and practiced leads to incremental changes or improvements. To get transformational change, innovation in this sense will not be sufficient. We need a more encompassing approach that will lead to revaluation, re-design and re-positioning of agricultural activities. We reserve the term 'systems innovation' for that approach. So, paraphrasing on the description by Porter, systems innovation leads to new ways of doing things that are commercialized and lead to new and improved configurations with the surrounding physical and social environment. This constitutes the desired transformational change that will 'turn new ideas into cash, environmental efficiency and social justice'. And for that reason a difficult trick has to be performed: not only innovate the agricultural production system, but at the same time innovate the way in which it performs innovation.

1.3. Sustainable development as an innovation target

The changing relationship between agriculture and society makes sustainability values more important in the agricultural sector. Government, citizens and consumers criticize agriculture for the negative side effects of agricultural production. The shift to a more sustainable agricultural sector is crucial to comply with these changing values. We need to shift to a new green(er) revolution. As Steiner puts it in relation to the 2009 UNEP report on the environmental food crisis: "*We need a Green revolution in a Green Economy but one with a capital G*" (UNEP 2009). Unilever's Patrick Cescau stated in a keynote speech at Columbia University in 2007 that sustainability and corporate social responsibility have become the main drivers for food industry. For that reason, food industry is not in a position to *not* focus on sustainability aspects of food production.

Sustainable development requires thinking and acting from a triple bottom line-perspective, focusing on economic prosperity, environmental quality and social justice (Elkington 1999), also often referred to as profit, planet and people aspects. This notion of the triple P-values (people, planet and profit) has given rise to a huge debate on the exact definition of sustainability (WRR 1995). In this book we will not go into this discussion. We use sustainability and sustainable development from a triple bottom line perspective, in which profit is used in the sense of measurable (economic) quantity, planet as environmental efficiency and people as relevant values of the physical and social environment, such as labor condition, animal welfare, and accessibility to sustainable food.

These triple P-values are often perceived by producers as complex factors to take into account. Sustainability, however, is not a difficulty that food industry and other executives should consider a burden. According to Nidumolu et al. (2009) sustainability should be industry's focal point for innovation. Not only in terms of social corporate responsibility, such as Cescau argues, but also from a commercial point of view, businesses should shift their innovation target to sustainable development. Environmental friendly production can lower resource use and therewith lower the production costs. More important though, is the perspective for more sustainable innovation. The competitive landscape, as Nidumolu et al. argue, is being transformed by the quest for sustainability. Early movers, whose focus on innovation is sustainable development, will have an advantage over other businesses, as sustainability will be a fundamental aspect of development, now and in the future. Besides being a way of ensuring cost reduction, becoming environmental friendly through a process of innovation aimed at sustainable development may also generate additional revenues, because environmental friendly products will perform better or the innovation enables the creation of new businesses.

The operational approach for industry to deal with sustainable development is given by the three licenses indicated by Casimir and Dutilh (2003): the license to produce, license to sell and license to operate (see Figure 1.3).

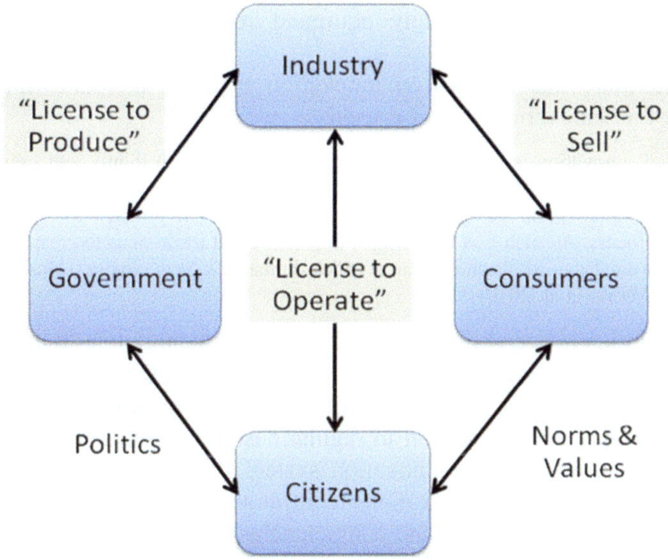

Fig. 1.3. The three licenses needed by industry (Casimir and Dutilh 2003).

Complying with the three licenses is crucial for industry to survive. Attention has been primarily given to the license to sell by delivering good quality products and services for a competitive price. The license to operate for agriculture is debated in the public domain, because here the negative side effects of agriculture to the environment and to social values are debated. As a result, the license to produce is constricted by additional rules and regulations that agriculture is faced with. For the agricultural sector this implies that the license to operate should be regained in order to keep the licenses to produce and to sell. Innovation towards sustainable development means that these three licenses should be kept in mind while innovation processes are initiated.

Hence, for reasons of profit, planet and people values, we need an innovation system that enhances innovation towards a more sustainable development. As the former paragraph argued, 'doing better things' requires a re-design or transformation of the innovation system. And the main design criterion for this transformation is sustainable development. In the next paragraphs we discuss the characteristics of the innovation system and come up with several ideas on how such a transformation should be approached.

1.4. Characteristics of innovation and the innovation system

How could an innovation system that contributes to sustainable development look like? Before diving into aspects of such an innovation system, we will first describe the characteristics of the innovation system in more detail to identify what aspects in the system are not adequately equipped for contributing to sustainable development.

In the literature that focuses on innovation systems, the following working hypothesis of an innovation system is used:

> A system of innovation is that set of distinct institutions which jointly and individually contributes to the development and diffusion of new technologies and which provides the framework within which governments form and implement policies to influence the innovation process. As such it is a system of interconnected institutions to create, store and transfer the knowledge, skills and artefacts which define new technologies. (Metcalfe 1995: 38, in: Van Mierlo et al. 2010)

This description puts emphasis on the diffusion of technologies within the innovation system. However, we have already described that the type of innovation we are after requires not only innovation in technology, but also in skills and competencies of people and innovation in regulations and institutional arrangements. So, applied to an agricultural innovation system this would imply a system in which multiple actors and institutions contribute to the development of new technologies ('hardware'), skills and competencies ('software') and regulation and institutions ('orgware') and the diffusion thereof in the agricultural sector. To that end, a framework is needed that supports an interactive, non-linear innovation process in which shared management of technologies, knowledge, skills and institutions can flourish. Such an agro innovation system will support the sustainable development of the agricultural sector by constantly producing 'better things' in the agricultural arena.

Leeuwis et al. (2006) identify defaults in the current knowledge infrastructure that obstruct a fruitful interaction of different disciplines. And it precisely this type of interaction that is desperately needed to stimulate innovations that will bring about transformational change that will contribute to sustainable development. One example concerns the specialization of agricultural production processes. A great achievement of the green revolution was the improvement of crops. These improvements were a result of extensive research in a number of specialized research centers aimed at specific crops. This specialized research, for example the rice research at IRRI (International Rice Research Institute) based in The Philippines, has led to a whole range of new varieties that dramatically improved rice yields. So, specialization has led to the successes of the agricultural sector, but it fragmented the research community into highly specialized groups that are becoming increasingly unable to collectively address complex problems and challenges that we face when dealing with sustainable development. Even further improving a variety will not lead to transformational change. For that to happen, we need to reflect on the environmental and social problems and constraints that are coupled with, again for example, rice growing in different agro-ecological and social conditions. Limited transparency and access to other knowledge frustrates the possibility to address the multiple facets of sustainable developments in a coherent fashion.

In innovation sciences, innovation is nowadays perceived as 'an interactive, non-linear process in which multiple actors (e.g. firms, research institutes, intermediaries, customers, authorities, financial organizations) depend on each other in realizing innovation' (Van Mierlo et al. 2010: 320). There is a growing attention in innovation sciences for the functions in an innovation system that have to be served: entrepreneurial activities, knowledge development (learning by searching and learning by doing), knowledge diffusion through networks, guidance of the search, market formation, resources mobilization and creation of legitimacy (Hekkert et al. 2007). This variety of aspects and functions indicate that the complexity of innovation has become clear in the discipline of innovation studies.

Innovation policies, however, are still rather linear. They still appear to be knowledge driven, as if innovation just depends on creating knowledge, rather than applying this knowledge in practice (Smits and Kuhlmann 2004). Especially in day-to-day innovation policies and innovation stimulation programs, a clear market or pull perspective is lacking. Innovation policies and programs are measured by the number of scientific papers that have been produced or by the 'strategic' knowledge that was developed. But the real performance indicator for innovation is its societal impact. Did the innovation lead to commercialization? And did it improve the efficiency with respect to resource use, or other non-renewables? And what about its impact on social values like labor conditions or animal welfare? By omitting these crucial aspects of innovation, we are trapped in generating numerous potential ideas that mostly lack a clear trajectory of implementation. In our experience, most orchestrated innovation efforts lead to the development of concepts, but a concise risk assessment is generally lacking as is an investment strategy that takes into account these risks. Since innovation at the end of the day is all about 'turning ideas into cash' (or other values, for that matter) the innovation as such – applying knowledge in practice – does not easily occur.

All together, one can state that the current knowledge infrastructure is very well capable of developing new specialized knowledge for successful agricultural production, but it fails to contribute to applying this knowledge in the process of stimulating a more sustainable development in the agricultural sector. Or more precise: the knowledge system is very strong in producing knowledge, but fails to realize innovation. This is sometimes referred to as the 'innovation paradox'. In order to enhance innovation towards sustainable agricultural development, we thus need an agricultural innovation system that applies the generated knowledge not only for productivity increase, but also makes use of this knowledge in a meaningful way to respond to the concerns that are raised by citizens, governments and consumers on environmental and social values.

This leads to the central question addressed in this book: What is needed to transform a linear innovation approach based on the assumption that the existing knowledge infrastructure generates the necessary knowledge, to a new agro innovation system in which meaningful knowledge is generated by multiple sources and that provides a real contribution to a more sustainable development?

We think that the answer to this question is to be found by actively organizing engagement between relevant stakeholders, including the knowledge – or academic – community.

1.5. System innovation as a process of engagement

Looking at innovation for sustainable development as a continuous process of engagement may prove to be a very promising perspective for realizing innovation. Engagement implies a respectful consideration of values that other parties bring to the table. This is precisely what is needed in the sustainability debate, since the arguments here revolve around the specific notion that is attributed to sustainability by different individuals, groups and institutions. Not only there will be differences of opinion on the relevance of profit against people and planet, but also within these broad categories different groups and individuals will have different ideas on the most pressing issues to be resolved.

The real boost for addressing the process aspects of sustainable development stems from the notion of 'wicked problems' and the recognition that sustainability itself can be considered as such. The notion of wicked problems evolved primarily in operations research and social planning as a reflection to the situation where a problem did not seem to have a clear solution. Rittel and Webber (1973) use nine more or less formal characteristics to discern between wicked and normal problems in the social planning domain:

1. There is no definitive formulation of the problem;
2. There is no stopping rule in problem solving;
3. Solutions are never true of false, but only better or worse;
4. There is no immediate or ultimate test of any given solution;
5. Every attempt for a solution counts significantly, there is no trial-and-error;
6. There is not a set of potential solutions nor permissible operations;
7. Every problem is essentially unique;
8. Every problem can be considered a symptom of another wicked problem;
9. The explanation of the problem determines the proposed solution.

If we take these characteristics and hold them against the problem of how to attain sustainability, all of them seem to apply. The ongoing debates and vast body of academic and polemic literature shows that there is an ever-expanding list of definitions of sustainable (1). And the same literature reveals that without a clear or agreed definition, there is also no way of deciding whether or not the problem is tackled (2). The nature of the discourse shows that the argumentation for the different definitions depends heavily on differing value sets that are used. So, there is no way of discerning between 'true' or 'false' solutions and instead an endless discussion on 'better' or 'worse' as perceived by the different stakeholders is unfolding (3).

Since all solutions are under constant debate and the outcomes are part of discussion, an ultimate test for any solution is impossible (4). And if any solution has any backing at all, it readily becomes part of experimental programs or projects and thus forms a significant solution as part of a larger effort (5). Since the normative or value aspects drive the problem definition and therefore the solution, there is no way of telling whether a given solution is not operational or does not fall within a set of potential solutions (6). And for the same reasons – the problem perception and solution driven by underlying values – every problem is in essence looked upon as unique. Not so much because of its observable traits, but much more because the problem as perceived is unique (7). Now, part of the ongoing debate is caused by the fact that a proposed solution by a certain stakeholder will give rise to the counter-argument by another stakeholder that their problem will not be solved at all. This intrinsic polarization of the ultimate solution sees to it that indeed any solution will give rise to another problem that will prove to be 'unsolvable' (8). And finally, since the problem definition is based on the problem as perceived by the various stakeholders, all solutions that are derived from such a problem definition will be determined by the same underlying value set (9).

So, it appears that sustainability issues can be regarded as wicked problems. Stressing the importance of solving the related problems will therefore not lead to satisfactory solutions. Instead, accepting the wicked character of the problem and address it in a process of engagement may be more useful. In such a process, different stakeholders can bring their values and related problem perception to the table. In recent literature, various authors have suggested this approach and came forward with elements of a possible action perspective. The analytical bridge between the notion of wickedness and the different approaches that rely on non-linearity and complexity is formed by the uncertainties that come into play once wickedness of the problem is recognized. For example Hjorth and Bagheri (2006) identify that we need non-linear and organic systems thinking to cope with complex and self-organizing systems. Palmer et al. (2007) advocate a trans-disciplinary approach that includes higher order thinking that transcends discipline boundaries and the generation of new knowledge and new resolutions not available in multidisciplinary and interdisciplinary environments.

Van Bueren et al. (2003) bring forward the notion that three different types of uncertainty play a role in the wicked behavior of problems:

1. Cognitive uncertainty: we just do not know enough about the problem and the potential causes to effectively address the problem.
2. Strategic uncertainty: there are many actors involved that have different perceptions of the problem and its solutions. Diverging and sometimes even conflicting strategies are the results that cause stagnation in the debate, or – in rare cases – lead to surprising and unexpected outcomes.
3. Institutional uncertainty: decisions are made in different places and policy arenas in which actors from various policy networks participate. This gives rise to a highly fragmented institutional setting that breaks down the legitimacy of choices and actions.

However, to deal with these uncertainties some mode of operation is needed that takes up these uncertainties in a process-architecture that uses them as a practical guidelines for dealing with these intrinsic properties of sustainability. A first step into that direction we find in the analysis of Norton (2005) who proposes an adaptive ecosystem management approach. He draws heavily on the pragmatists' philosophical worldview that puts experiments at centre stage. He proposes to deal better with sustainability (and thus wickedness of the problem) by adhering to three principles:

1. Experimentation

   Since there is no single pathway possible to a solution and as a result of the value driven problem perception, the only way forward is by executing experiments that will reveal whether any proposed solution or approach will improve the current situation. In Norton's view, all knowledge related to a given problem must therefore be tested through experimentation. This takes away the up-front distinction between facts and values and brings everything back to a practical assessment whether or not a proposed partial solution or approach does effectively work.

2. Multi-scalar analysis

   Norton suggests that especially the influence of different spatial and temporal scales introduce a lot of value driven assumptions related to the problem and consequently the solution to the table. So, in his view these different levels of scale must be addressed when handling the problem, with special attention to communal values that are relevant to intergenerational issues and economic values that are relevant for short-term effects.

3. Place sensitivity

   The second largest cause of disagreement on problem definition and potential solutions, again according to Norton, can be found in the local attributes that are attached by the various stakeholders to the challenge. The basic fact that all approaches or intervention to improve sustainability issues must be specified in a given location underlies this notion. Only if the location specific problem description is made clear, Norton feels that there is a real opportunity to describe the challenge at hand and thus arrive at feasible modes of operating on the situation.

Although these first steps from analysis to action perspective shed some light on how to go about, in essence Norton's advice is still pretty vague. Of course, the notions of experimentation, of multiple scales and of relativity to location are all relevant, but so is the need for interaction and collective learning to understand and deal with the different types of uncertainty that were described by Van Bueren et al. (2003). And also the trans-disciplinary approach proposed by Palmer et al. (2007) that will lead to the necessary new knowledge. Based these notions and our initial experiences in the TransForum program (see Appendix I for more information on the program), we conclude that Norton's recommendations lack an 'actor element'. Dealing with sustainability is also a social learning process in which different individuals and (social) groups are involved. In this type of learning processes there is no pre-conditioned route, but the interaction of different perception of different stakeholders to any given challenge results in a learning trajectory (Van Mierlo et al. 2010).

Another aspects that is still lacking, is the role of science in this inherently *social* learning process. If the input of all stakeholders in a social learning process is relevant, then the traditional role of science and academia as the guardian of the truth is contested. For example: Ludwig (2001) explains that complexity and wickedness of problems indicate the end of the traditional management paradigm. He further states that the lack of description of the problem and its potential solutions leads to a different approach to the issue of problem solving and to a debate on the role of science and scientific analysis in this context. So we see that science no longer can play the role of an un-challenged arbiter. Science and scientists are forced to share their role as professional adviser to the decision-making process with a larger group of involved stakeholders. This forms an essential element of the needed engagement of science. Add to this the same levels of engagement in business, government and societal groups and the outlines of a process approach to deal with sustainable development emerges.

1.6. Five motivating assumptions for enhancing sustainable development in agriculture

Dealing with wicked problems and thus dealing with sustainable development needs experimentation (Norton 2005). The TransForum program was initiated to gain experience with this type of experimentation in the agricultural sector (see Appendix I). Our practical experiences in a number of action experiments in the agricultural sector reveal that there is not one recipe for bringing about the transition that is searched for. However, based on the notions in literature (see Section 1.5) and experiences in the first year of the program we developed a set of motivating assumptions as a guideline for our work. The set of motivating assumptions provides a framework that we used to shape all following action experiments. These assumptions can be seen as inductive hypotheses that we try to validate based on further experiences in the rest of the program. This approach is based on the methodology of grounded theory (Strauss and Corbin 1990) that describes the way in which a large number of findings can be used as a database on which inductive theories are built. So, instead of first devising a theoretical analysis, we started with experimenting and induced theory from our findings. This led to the following assumptions that together constitute a sequential line of reasoning:

1. Sustainable development is a dynamic process
2. Sustainable development needs system innovation
3. System innovation is a non-linear learning process

4. System innovation requires a multi-stakeholder approach
5. Multi-stakeholder approaches imply trans-disciplinary knowledge creation

We will discuss them one by one.

Assumption 1: Sustainable development is a dynamic process

Since the notion of sustainable development is engrained in the various value systems that different stakeholders bring to the table, an exact definition of sustainability may seem desirable but is impossible. Here we see a clear resemblance to notions like 'health' and 'happiness'. All these concepts cannot be definitely defined, but still they play a crucial part in human development. So, if we treat sustainable development the same way as these other concepts, a dynamic perspective emerges. A constant effort into a direction of more sustainability (indeed: sustainable development) is in that sense comparable to a constant effort to improve health and happiness. This shows that treating sustainability as a dynamic process does not lower the ambition; it may well enforce the ambition by stressing the need for a constant flow of activities, research, entrepreneurial efforts and experiments into the direction of a more sustainable development.

Assumption 2: Sustainable development needs system innovation

Sustainable development requires a transformational change in the agricultural arena. We have to redesign the way in which agriculture is being performed. The transformed agricultural production system must be able to not only efficiently produce from a monetary perspective, it must also meet the demands that stem from environmental constraints and it must comply with demands that are related to aspects such as animal welfare and working conditions. This can only be achieved with a transformational change of the production system. The basis for this is a technical innovation that opens the windows for a better way of production. But the application of new modes of production also needs adaptations in skills and competencies of the professionals involved, so a social innovation. And, moreover, the new production system must also fit into the organizational context. This will only be possible if not only the production system itself is transformed, but also the context in which it must operate. This calls for an innovation in the production system, its participants and its surrounding organizational environment. Such a transformational change is denoted as a systems innovation: an innovation that does not only shifts the mode of production, but also the system in which it must take place. So, systems innovation is a prerequisite for a shift into a direction of more sustainable development in the agricultural domain.

Assumption 3: System innovation is a non-linear learning process

A dynamic innovation process is required to enable the technical, societal and organizational changes that system innovation calls for. The required dynamics in this innovation process are created by reflexive feedback loops (Carlsson et al. 2002). Learning as a feedback in the innovation processes is crucial (Gibbons et al. 1994). This includes learning about technical aspects, about skills and competencies of the actors involved and about organizational aspects of the realization of the innovation. This type of learning is referred to by Gibbons et al. (1994) as Mode-2: knowledge that is not transferred from science to society, but performed to contribute to the application (Regeer 2010). Science and society both have a role in producing and utilizing knowledge for complex change processes such as system innovations. This all implies that a transformation of the innovation system must take into account the reflexive cyclical learning processes that are needed for a more sustainable development. This can be achieved by organizing feedback loops between the different stages in the innovation process and the different actors.

Assumption 4: System innovation requires a multi-stakeholder approach

A non-linear learning process implies input from different actors: science as well as society. A system innovation towards a sustainable development process of the agricultural sector is a responsibility of all actors in the sector. Farmers, citizens, policy-makers, researchers and governments all have values concerning environmental, social and economic aspects of agricultural production. Relevant stakeholders must be included in a process of engagement to strengthen the license to sell, operate and produce. As a minimum requirement we propose to invite at least stakeholders from four different domains at the table: knowledge institutes, entrepreneurs, societal organizations and governments. The non-linear learning process of system innovation thus requires a multi-stakeholder approach in order to successfully contribute to sustainable development.

Assumption 5: Multi-stakeholder approaches imply trans-disciplinary knowledge creation

The current specialization of knowledge creation in the agricultural domain puts heavy constraints on the needed interactive processes for system innovation. The changing role of science and society in system innovation calls for a new mode of working in which the knowledge of all participants is utilized in a novel way. Johansson (2004) calls this 'the intersection', a situation where actors with different disciplines and experiences come together so that new ideas and thoughts are created. In the interaction that is brought about by a multi-stakeholder learning process, theory and practice and different disciplines are combined. This transdisciplinary approach creates the new knowledge that is needed for system innovations towards a more sustainable development in agriculture (Peterson 2009).

1.7. Stimulating sustainable development in a learning-by-doing experiment

This set of five motivating assumptions are applied as a framework to stimulate sustainable development in a diverse set of over 30 action experiments (see Appendices II and III for more information and an overview of the action experiments). The action experiments were all aimed at improving the sustainability performance of (revised) agricultural activities. By gaining experience in action experiments, we wanted to (1) test the validity of the assumptions in a number of very different situations, and (2) get some insight in the practical do's and don'ts when applying the framework. Ultimately, the assumptions together with the more practical implications in terms of do's and don'ts together constitute the groundwork for taking next steps in transforming agro innovation to a system in which the innovation process will have an actual contribution to sustainable development.

During the execution of the action experiments some persistent issues appeared that hamper the application of the assumptions and therewith realization of the actual innovation. For example, it appeared that the diversity in images of sustainable development can be a difficulty in action experiments that aim to realize a more sustainable development. Also the rigidity of new technical inventions in realizing innovation processes came out as a challenge. And what are institutional blockages in realizing innovation and transition? How are more sustainably produced products perceived by the consumer? What role does the market (have to) play in that process? As most processes around wicked problems, the questions that were articulated in the various experiments are not straightforward to answer. Most of these issues require a more fundamental analysis in order to work with the motivating assumptions proposed. Appendix I gives more information on the relation between the action experiments and the scientific research in our approach.

This book is aimed at analysis and clarification of some though problems that we came across in applying the five motivating assumptions in the action experiments. This challenge is approached from different scientific disciplines.

In the next five chapters we will explore key issues that were raised in the action experiments and that influence the dynamics of sustainable development and system innovation in the agricultural sector. This will increase our understanding of the innovation process and open up action perspectives in this process. By the approach we use, using grounded theory as a starting point for experimentation and thoroughly analysis persistent issues out of these experiments, we show the role of science in this social learning process and how scientific research can contribute to applying new fundamental knowledge for sustainable development in practice.

Chapter 2 is about inventions related to sustainable development of agriculture. Most innovations start with an invention. Inventions are crucial for bringing about innovative and more sustainable processes in agriculture. But technical inventions alone do not necessarily create an innovation. This chapter addresses the leap changes that are required to bring about innovation and ways in which inventions and innovation can be organized.

Chapter 3 considers the organization of innovation and transition in agriculture. How can one ensure that the inventive energy for innovations in sustainable development of agriculture and green space can be used to be put into effect? Following on from this, how can individual innovations lead to transition through systemic innovations?

Chapter 4 deals with the dominant role of images for the sustainable development of agriculture and green space. Sustainability is a goal that is hard to quantify. To the question 'what is sustainable development?' can be responded to with different images of sustainability. These images have a strong influence on the outcome of an innovation process.

Chapter 5 studies how the consumption of more sustainably produced food can be promoted, and how the demand side of more sustainable production processes can support a sound business model. How do individuals perceive sustainability when they are in the supermarket about to choose their evening dinner? For sustainable innovations in the agricultural sector to be successful, the (quantitative or qualitative) added value of sustainably produced products must be perceived as such by consumers.

Chapter 6 researches the effectiveness and validity of the set motivating assumptions in the action experiments. It also gives insight on how to organize learning processes for sustainable development? How do multi-stakeholder project teams work together? How does trans-disciplinary knowledge contribute to the innovation process?

In *Chapter 7* concluding remarks will be made about what the chapters learn us about what is needed to shift to a new agro innovation system in which meaningful knowledge is generated by multiple sources and that provides a real contribution to a more sustainable development. The research revealed in the individual chapters give insights from different scientific disciplines for the set of motivating assumptions to enhance innovation for a more sustainable development in the agricultural sector.

1.8. References

Beddington, J. (2010). Food security: Contributions from science to a new and greener revolution. *Philosophical Transactions of the Royal Society of London B Biological Sciences*, *365*(1537), 61–71.

Bekman, H. (1998). Controlled quality meat: The Dutch experience. *Journal of Food Safety, 18*(4), 363–370.

Boogaard, B.K., Bock, B.B., Oosting, S.J., Wiskerke, J.S.C. and van der Zijpp, A.J. (2010). Social acceptance of dairy farming: The ambivalence between the two faces of modernity. *Journal of Agricultural and Environmental Ethics*, doi: 10.1007/s10806-010-9256-4.

Carlsson, B., Jacobsson, S., Holmén, M. and Rickne, A. (2002). Innovation systems: Analytical and methodological issues. *Research Policy, 31*(2), 233–245.

Casimir, G. and Dutilh, C. (2003). Sustainability: A gender studies perspective. *International Journal of Consumer Studies, 27*(4), 316–325.

Dunn, E.C. (2003). Trojan pig: Paradoxes of food safety regulation. *Environment and Planning A, 35*(8), 1493–1511.

Elkington, J. (1999). *Cannibals with forks: The triple bottom line of 21st century business.* Oxford: Capstone Publishing Ltd.

FAO (2002), *World Agriculture: Towards 2015/2030.* Food and Agriculture Organization of the United Nations, Rome.

Gibbons, M., Limoges, C., Nowotny, H., Schwartzman, S., Scott, P. and Trow, M. (1994). *The new production of knowledge: The dynamics of science and research in contemporary societies.* London: Sage.

Hekkert, M.P., Suurs, R.A.A., Negro, S.O., Kuhlmann, S. and Smits, R. (2007). Functions of innovation systems: A new approach for analysing technological change. *Technological Forecasting & Social Change, 74*(4), 413–432.

Hjorth, P. and Bagheri, A. (2006). Navigating towards sustainable development: A system dynamics approach. *Futures, 38*(1), 74–92.

Johansson, F. (2004). *The Medici effect: What elephants and epidemics can teach us about innovation.* Boston, MA: Harvard Business Press.

Jongbloed, A.W. and Lenis, N.P. (1998). Environmental Concerns about Animal Manure. *Journal of Animal Science, 76*(10), 2641–2648.

Leeuwis, C., Smits, R., Grin, J., Klerkx, L.W.A., van Mierlo, B.C. and Kuipers, A. (2006). Equivocations on the post privatization dynamics in agricultural innovation systems. In *The design of an innovation enhancing environment. Working papers no. 4* (pp. 3–58). Zoetermeer: TransForum Agro & Groen.

Ludwig, D. (2001). The era of management is over. *Ecosystems, 4*(4), 758–764.

Nidumolu, R., Prahalad, C.K. and Rangaswami, M.R. (2009). Why sustainability is now the key driver of innovation. *Harvard Business Review, 88*(9), 1–10.

Norton, B. G. (2005). *Sustainability: A philosophy of adaptive ecosystem management.* Chicago: University of Chicago Press.

Palmer, J., Smith, T., Willetts, J. and Mitchell, C. (2007). Creativity, ethics and transformation: Key factors in a transdisciplinary application of systems methodology to resolving wicked problems in sustainability. In: *Systemic development: Local solutions in a global environment.* Auckland, New Zealand: Institute for Sustainable Futures, University of Technology.

Peterson, H.C. (2009). Transformational supply chains and the 'wicked problem' of sustainability: Aligning knowledge, innovation, entrepreneurship, and leadership. *Journal on Chain and Network Science, 9*(2), 71–82.

Porter, M.E. (1998). *On competition.* Boston, MA: Harvard Business Press.

Regeer, B.J. (2010). *Making the invisible visible. Analysing the development of strategies and changes in knowledge production to deal with persistent problems in sustainable development.* Oisterwijk: Boxpress.

Rittel, H.W.J. and Webber, M.M. (1973). Dilemmas in a general theory of planning. *Policy Sciences, 4*(2), 155–169.

Schiere, J.B. and Van Keulen, H. (1999). Rethinking high input systems of livestock production: A case study of nitrogen emissions in Dutch dairy farming. *Tropical Grasslands, 33*(1), 1–10.

Smits, R. and Kuhlmann, S. (2004). The rise of systemic instruments in innovation policy. *International Journal of Foresight and Innovation Policy, 1*(1/2), 4–32.

Strauss, A.L. and Corbin, J. (1990). *Basics of qualitative research: Grounded theory procedures and techniques.* Newbury Park, CA: Sage publications.

UNEP (2009). *The environmental food crisis: The environment's role in averting future food crises: A UNEP rapid response assessment.* Arendal, Norway.

Van Bueren, E.M., Klijn, E.H. and Koppenjan, J.F.M. (2003). Dealing with wicked problems in networks: Analyzing an environmental debate from a network perspective. *Journal of Public Administration Research and Theory, 13*(2), 193–213.

Van Latesteijn, H.C. (1998). Future land use in Europe. In E.M. Barron and I. Nielsen (eds.), *Agriculture and sustainable land use in Europe: Papers from conferences of European environmental advisory councils* (pp. 101–115) (illustrated ed.). The Hague; Boston: Kluwer Law International.

Van Mierlo, B., Leeuwis, C., Smits, R. and Woolthuis, R.K. (2010). Learning towards system innovation: Evaluating a systemic instrument. *Technological Forecasting & Social Change, 77*(2), 318–334.

Wallinga, D. (2009). Today's food system: How healthy is it? *Journal of Hunger and Environmental Nutrition, 4*(3–4), 251–281.

WRR (1995). Sustained risks: A lasting phenomenon, report no. 44. The Hague: SDU.

Chapter 2. Inventions for future sustainable development in agriculture

Evert Jacobsen[1], Pieter J. Beers[2] and Arnout R. H. Fischer[3]

[1]Evert.Jacobsen@wur.nl; TransForum and Plant Breeding, Wageningen University, Wageningen, The Netherlands

[2]PJ.Beers@wur.nl; TransForum and Education and Competence Studies, Wageningen University, The Netherlands

[3]Arnout.Fischer@wur.nl; TransForum and Marketing and Consumer Behaviour, Wageningen University, Wageningen, The Netherlands

Abstract This chapter is directed to the importance of different inventions as driver for sustainable development of agriculture. Inventions are defined as radical new ideas, perspectives and technologies that hold the potential to trigger a change in sustainable agriculture. Innovation is based on one or more inventions that can bring a major breakthrough or have a smaller impact. Inventions can be science driven or society driven. They have been crucial in history of the existing agricultural knowledge infrastructure and will remain important in the future as driver for sustainable development. The main difference with the past is that inventions and their implementation will not always follow the old linear process of fundamental – (university), strategic – (research institute) and applied research with extension – (applied institutes) but that these processes are often non-linear and more complex, including society. For inventions to become successful innovations and to prevent negative side effects not only technical hardware elements of an invention are important but also software and orgware elements. Another important prerequisite for innovation through inventions is value capturing through entrepreneurship. Nowadays in agricultural life sciences patent rights are introduced which have to co-exist with, for example, plant breeders' rights in plant varieties with its principle of breeders exemption in order to preserve open innovation. Lessons learned show that new solutions aiming at sustainable development can be organized by intersectional inventions, directional inventions or open innovations sometimes followed by a switch between them.

2.1. Inventions – an introduction

An invention is the result of a highly creative process with outcomes beyond what is currently known. Inventions are initial ideas that may either require further elaboration, development or analysis to assess their true potential. After this, they

H.C. van Latesteijn and K. Andeweg (eds.), *The TransForum Model: Transforming Agro Innovation Toward Sustainable Development*, DOI 10.1007/978-90-481-9781-1_2,

can become the input around which innovations for sustainable agriculture are organized. In this chapter we define inventions as radical new ideas, perspectives and technologies that hold the potential to trigger a change. In this context, an invention is 'creating something that did not exist before' leading to new ideas for products and processes and an innovation is 'thinking creatively about something that already exists' (Branscomb and Auerswald 2002). An invention becomes an innovation when it is widely implemented and used in society. Recent inventions that have led to innovations are the Internet, e-mail, mobile telephone, nano-technology, biotechnology, and the concept of energy producing greenhouses.

Inventions play an important role in all sectors of our society, including in the agricultural sector. As they have played an important role in the development of our agricultural system in the past, inventions can and should be of importance as a driver for innovations towards a sustainable development of agriculture in the future (Leeuwis et al. 2006). However, as already addressed in Chapter 1, society has become an important stakeholder in development processes. Nowadays inventions are much easier debated in society than in the past (Smits 2002). Such discussions in society have focused on the benefits of the inventions, but can also constrain the use of a new invention, therewith limiting its potential to become an innovation. The main reason for this change is the increased awareness of society for all kinds of potential negative side effects of inventions and their implemented innovations for environment and health.

For innovation towards the sustainable development of the agricultural sector, it is important to know what positively or negatively influences inventions and how inventions can be positively directed in the future. Thus the questions are: What aspects are of importance to turn inventions into successful innovations? Is it possible to select inventions directly, including an understanding of their limiting conditions, in such a way that they improve sustainability?

This chapter aims to describe and discuss the most important elements of inventions for innovations in agriculture, and how the development and use of future inventions could be steered in such a way that they have a higher potential to turn into innovations. The chapter first discusses roles, types and aspects of past and modern inventions that influenced their success in Section 2.2. It proceeds in Section 2.3 with ways of stimulating inventions into an innovation using various examples of success. The chapter concludes in Section 2.4 with a short discussion on lessons learned and the expected future role of inventions for sustainability in the agro innovation system.

2.2. Inventions in agricultural infrastructure

In 1798 Thomas Malthus predicted that the world could not feed an exponentially increasing population. His prediction did not materialize. New inventions have been tremendously important to support sufficient agricultural production (see

Figure 1.2 in Chapter 1 for yield increase due to the green revolution). At this moment we stand for another major challenge. The world should be able to feed a doubled population by 2050 and at the same time do this in a more sustainable way. It is without any doubt that new inventions will be crucial to stimulate and enable a sustainable development in today's and future agricultural production. Inventions have helped before to allow for the required productivity increase and should now be aimed simultaneously at people, planet, and profit aspects.

To get a grasp of how inventions and their subsequent innovations are influenced, we discuss two main contextual factors that must be considered on how inventions can be implemented towards innovation for sustainable development. The first factor deals with the ownership of inventions and the different modes of how to handle that. The second factor deals with the conditions necessary to transform inventions into innovations. These factors are illustrated by inventions from the agricultural sciences.

2.2.1. Ownership issues: Inventions, innovations and IPR

2.2.1.1. Ownership and agricultural inventions of the past

The green revolution inventions were characterized by an open innovation structure. Inventions were not protected by Intellectual Property Rights (IPR) and, therefore, available for everybody. This improved food security in many countries significantly. New varieties with a higher yield could be used by everybody. Two major inventions of the green revolution are discussed here are (1) breeding of short straw cereal varieties and (2) hybrid varieties with modern plant production techniques transforming agriculture by using more fertilizers and pesticides[1] (Borlaugh 2000a, 2000b). Both are typical ways of a linear innovation where a scientific finding is transferred onto the market by a process of fundamental – (university), strategic – (research institute) and applied research with extension – (applied institutes) and limited societal influence.

The first successful examples of an open invention are 'dwarf mutants' of wheat and other cereals. Those new varieties have much shorter stalks because of a dwarf mutation (Salamini 2003; Fig 2.1). This invention enabled the development of modern agricultural technology in cereals. It had a major impact on global scale, mainly because a system of plant breeders' rights (PBR) protected the ownership of the new invented variety, without limiting others to use the new variety as breeding parent and improve on it. The foundation for these rights was drawn up by the International Union for the Protection of New Varieties of Plants (UPOV). The Convention was adopted in Paris in 1961 and revised in 1972, 1978

[1] Besides its major advantages, the green revolution has afterwards also been criticized as one of the first examples of being not sustainable and partly reducing agro biodiversity because of less use of locally adapted varieties.

and 1991 (for a more elaborate discussion on the reasoning behind the PBR as a 'sui generis' system being distinct from patent rights see Rimmer 2003 and Dubois 2001).

The system of plant breeders' rights allowed breeders to use UPOV protected varieties as crossing parent in breeding programs of competitors to create new varieties (breeders exemption). This means that all traits in the protected varieties can be used for free in any new variety. UPOV protects the genotype of the plant, but not the new trait or its genetic background. In this way, a new variety with a new trait can be replaced quickly by a newly improved variety of a competitor. As a result, short straw varieties could also directly be developed in developing countries using locally adapted plant material. This invention tripled cereal production since the 1960s, changing India into a wheat exporting country (Borlaugh 2000a, 2000b).

Fig. 2.1. The benefits of being short. A: Varieties from early 1900s were as tall as a normal person. B. modern short straw varieties have much shorter stalks because of a dwarf mutation (Salamini 2003).

A second important example is the concept of hybrid varieties. The concept of hybrid varieties was practiced, more than 90 years ago, in the USA in maize, vegetables and other (cereal) crops (Crow 1998), and was introduced in Europe more than 60 years ago (Marton et al. 2003). Hybrid varieties are based on crosses between two inbred lines with optimal combination ability for yield. Because of hybrid vigor, hybrid varieties increased yield up to six fold.

Integrated into the 'hybrid invention' is the natural breeders' seed protection. Due to the impossibility to multiply the same hybrid seeds on the hybrid plant itself the seed is intrinsically protected. This way of seed protection has triggered hybrid breeding and seed trading of many (vegetable) crops, also in countries with absence or weak implementation of UPOV breeders' rights, resulting in the establishment of a number of strong seed companies worldwide (Anonymous 2009a).

These two examples from past inventions show that plant breeders' rights have been a successful factor to stimulate new inventions. It is a way in which open innovation is ensured, while still protecting new varieties. In general it can be said that this open system of plant breeding added a yearly yield increase of 2–3%.

2.2.1.2. Ownership and modern inventions in agricultural sciences

One of the emerging fields in agricultural life sciences is (green) biotechnology like 'omics', including genomics, proteomics and metabolomics, and genetic modification of organisms (GMO). Gene technology related to agricultural sciences, such as gene transformation or genetic modification (GM), is of importance for plant breeding (Bevan 1984). It can for example increase resistance of crop plants to (a)biotic stress caused by climate change, or reduce chemical use in crop protection. This field is also called the 'gene revolution'. In contrast to the green revolution, the gene revolution is much less open and increasingly subjected to industrial patent system. One of the main differences between PBR and patent rights (PR) is breeders' exemption which is allowed within PBR but not within PR and also not when a variety has been protected by both PBR and PR, at the same time. This double protection is valid, for example, when single PR protected traits have been inserted into the plant or protected methods have been used. Originating from the chemical and pharmaceutical industry, PR has been extended to cloned genes (Cohen and Boyer 1980) and biotechnological procedures (Bevan 1984) in the last few decades.

This development in ownership has allowed private companies to license inventions. In the USA, the Bayh-Dole Act[2] gave universities and institutes ownership of discoveries by PR rather than making them freely available in the public domain (Zilberman et al. 2001). This possibility of licensing of PR has greatly improved the transfer and application of technologies from universities to the private sector, but there are some serious drawbacks. Exclusive licensing PR of an invention to a single private (breeding) company has the potential to limit the usage of the invention to the larger benefit of society by other companies during the protection period.

In the future inventions can play an important role in relation to sustainable development. To feed the world and to replace petrol-based chemicals and fossil fuels, the maximum potential yield of crops has to increase. Inventions can open doors for new developments. For example, in nature, several pathways for photosynthesis, like C3 and C4, are found. Their photosynthesis efficiency is different as, for example, observed in C3- and C4-crops (Monson and Moore 1989). Developing plant varieties with the most efficient photosynthesis mechanism would increase the energy production by the plant to the benefit of the yield. An invention that may support this is to increase efficiency of photosynthesis by molecular biotechnology.

---

[2] P.L. 96-517; The Bayh-Dole Act, 1980.

However, exclusive rights and use of those inventions is not desirable, because a broad and quick implementation of inventions that can serve sustainability is required. The common interest of society may be better served in other ways, such as non-voluntary or compulsory licensing of patents (Reichman and Hasenzahl 2003).

2.2.1.3. Open innovation under attack

The earlier described examples of yield increase by the introduction of dwarf mutants and hybrid varieties show how open innovation in plant breeding can benefit agricultural production and the agricultural sector as a whole (Borlaugh 2000a, 2000b). This open competition model ensures that none of the involved companies can block progress of societal importance. It also ensures that new traits become available worldwide quickly. Still, the competition model is very clear: only the variety is protected, not their traits or genes. Breeders know how they can earn money.

Although the introduction of PR in agricultural life sciences is considered to be very important, in plant breeding, it affects the open competition model ensured by the UPOV system when more varieties are introduced with GM traits, protected by PR (Jacobsen and Schouten 2009). The present interpretation of using PR in life sciences does not allow breeders' exemption. The result will be that breeders have free availability of new traits from normally bred varieties but no availability during the protection period of GM-traits from varieties with patent protected traits. Increasing reliance on patented GM-traits will frustrate the existing open breeding system. This PBR system has so far been ensuring a rapid and widespread availability of successful traits in a sufficient number of different genetic backgrounds worldwide and in the existence of a financially well performing private breeding sector.

An additional complication in the co-existence of PR and PBR in the near future is the expectation that the same traits will be available in classical PBR protected varieties and in PR protected GM-varieties. For example, it is expected that certain (resistance) genes will become available in classical varieties, and that the same trait with the same gene will be present in GM-varieties which are not freely available for breeding. The only, if any, difference between these varieties might be the location of the resistance gene in the plant genome. This will make the resulting PR discussion highly complex, because it is not clear whether the gene can be used as being the outcome of classical breeding, or whether it is protected by GM patents. This can result in increasing numbers of genes coding for important traits which are not freely available for variety breeding, causing a slow down of yield gain worldwide.

In sum, the co-existence of both PR (representing more closed innovations by patents rights) and PBR (representing more open innovation by plant breeders' rights) in one variety demonstrate that the rules and regulations related to:

1. PR will result in a slowdown of the rapidity of optimal variety development, because of the lack of free availability of important GM-based genetic variation, and;
2. PBR is not stimulating expensive investments that are needed for quick progress with important cloned traits because of aspects like breeders exemption and farmers privilege.

Besides their individual advantages, consequences of the co-existence of PR and PBR are negative in relation to the sustainable development of agriculture and world food production. This problem should be solved in a satisfactory way, which allows the beneficial aspects of both systems. One compromise could be to continue with both protection systems under the condition of breeders' exemption (Jacobsen and Schouten 2009), which requires the owner of a patent to agree with free license to breeders' exemption of newly introduced GM-traits and the breeder of the new variety to pay a reasonable license fee afterwards.

2.2.2. Organization of inventions and innovation

The way inventions are organized can greatly influence their potential for implementation to become an innovation. Many inventors still try to innovate linearly. Nowadays, inventions for sustainable development have to be seen in a complex interaction between human creativity, technology and the market place. They can no longer be considered as driven strictly by technology and basic research (Smits and Boon 2008). The increasing influence and involvement of society on the actual innovation based on an invention is important to recognize. Hence, it is important to know how the implementation of inventions can be made more smooth and successful. Below we discuss some aspects that are of importance for realizing actual innovation.

2.2.2.1. 'Ware' elements of inventions and entrepreneurship

Innovation is a complex process that takes place at the level of specific products, business and sectors as well as at the level of (inter)national communities (Smits 2002). Successful innovation is a fruitful combination of different elements from technical, societal and economic point of view. Inventions and the innovations connected to them should nowadays take into account different 'ware' elements (Leeuwis et al. 2006; Anonymous 2003):

- *Hardware* which relates to the material innovation;
- *Software* referring to the new skills and knowledge to use and implement the invention (including tacit knowledge); and

- *Orgware* referring to new organizational and institutional conditions, influencing development and functioning of innovations (Smits 2002). This also includes regulatory frameworks, modes of operation with regard to IPR, role and influence of public opinion, etc.

In case of a lack of balance between the different 'ware' elements, the application of an invention may be limited or may even lead to a complete blockage by society. Failure to adequately address the food-specific concerns in society has influenced the success of 'green' biotechnology for GM crops, whereas the 'red' (health) and 'white' (industrial) biotechnology are more accepted in society (Lelyveld 2009). Similarly, society is reluctant or even negative towards modern, highly-productive animal production systems (Ten Napel et al. 2006).

Two different lines of reasoning explain the misfit between technological possibilities and societal acceptance. It is hypothesized by some that natural sciences have made larger progress in understanding natural processes than social scientists in understanding social processes and behavior (Norman 2005). From this point of view, it seems that social sciences are lagging behind to keep up with knowledge on societal acceptance of technological developments. Against this, it can also be argued that it is the late involvement of social sciences and public participation which has resulted in a lack of understanding of contemporary societal requirements for important innovations. This led to mistakes by important stakeholders in the initial introduction of for example GM products, which has created a lastingly hostile atmosphere in European society towards these inventions.

To overcome hostility towards inventions like GM crop development and highly productive animal production systems, the polarized debate must be turned into a dialogue to look ahead. Psycho-societal solutions among the stakeholders have to be found, which may take considerable time (Termeer 2009). Orgware elements are important to incorporate in order to overcome this challenge.

Nowadays, basic research programs allocate part of their budget to social and ethical research, to incorporate specific knowledge about societal implementation of inventions. In this way, soft – and orgware elements are set on a comparable level as hardware elements of research. However, to get the right balance between the ware elements, budgets should not only be set side-by-side. In our view, an integration of the ware elements is needed for the right balance, meaning that the ware elements should be naturally implemented together from the beginning and implementation of the development.

Entrepreneurship may be the key to effectively combine the needed hardware, software and orgware elements of inventions into a driver for innovation. An entrepreneur converts a new idea (invention) into a successful business (innovation) (Druckner 1993), turning ideas into cash. Entrepreneurship is crucial to provide prove of principle and drive stages towards the market, making an invention into an innovation. Besides the three ware elements of an invention, commercialization of the invention via entrepreneurship is equally important, (see Figure 2.2; MacGill and Outhred 2004). This, of course, requires a market demand for the new product and/or services of the invention.[3]

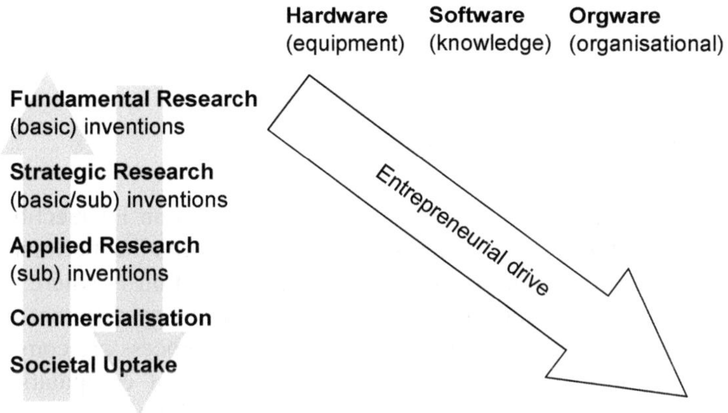

Fig. 2.2. 'Ware' elements of inventions obtained in research and their role in technical innovations and commercialization by entrepreneurship.

This section has described several aspects that are important for inventions and to turn an invention into an innovation. In sum, open innovation with protection of plant breeders' rights can boost new inventions that are needed for the sustainable development of the agricultural sector. Therefore, co-existence problems between PR and PBR must be solved. Additionally, to realize a successful innovation based on an invention, one should integrate hard,- soft- and orgware elements and ensure the commercialization of the invention.

2.3. Three ways to stimulate needed inventions

Since the need for a more sustainable development in agriculture calls for specific improvements on the performance of agriculture with respect to people, planet and profit aspects, it should be possible to stimulate inventions into those directions. So, instead of waiting for new inventions to occur as a result of serendipity, the question arises whether it is possible to select inventions directly, including an understanding of their limiting conditions, in such a way that they improve sustainability. Based on our experiences and recent publications on organizing innovations, we see three different ways of stimulating the needed inventions:

[3] In this chapter we will not go further into this issue, see Chapter 5 for an in depth discussion on the organization of market demand in sustainable production chains.

1. Organize so-called 'intersections' that address the challenges
2. Stimulate directional inventions
3. Enhance the process of open innovation.

This section illustrates these three ways with examples.

2.3.1. Intersectional inventions

Intersectional inventions are inventions originating from an intersection between different fields of expertise (Johansson 2004). This can be between different cultures or between realms of knowledge. The intersection is where ideas from different cultures, disciplines and fields meet causing many different inventions. At present the most important factors involved in this process are convergence of sciences, the movement of people, and the explosion of computation. Recent examples of such creative mash ups of knowledge are high tech applications such as Google Maps and Google Earth, e-commerce, biotechnology, nanotechnology but also organic farming. An important aspect of intersectional inventions is that they often need new or adaption of existing legislation, which introduces societal debate into the new issue (Yanarella 1975).

2.3.1.1. Organic agriculture

The invention of organic agriculture started in a more general debate with many disciplines involved, within as well as outside the agricultural sciences (ecology, nature, human health, philosophy, ethics and religion). The development of organic agriculture is taking place in a multi-stakeholder setting and orgware elements were addressed from the start of the invention. As a result, sufficient flexibility in the regulatory system has been created by (provisional) derogations to overcome specific problematic hardware elements, such as the use of organic manure, residue issues of pesticides and availability of organic seeds. Derogation to certain aspects of the regulation were granted and accepted without negative reactions from society (Jacobsen and Schouten 2009). The organic agriculture trend is mainly driven by society and is based on a philosophy highlighting the intrinsic value of nature. Organic agriculture has created a niche market that provides a societal accepted and conceived form of sustainable agriculture.

Organic agriculture is a good example of an agricultural production system where an orgware and software driven approach has taken precedence, and where the strong position in relation to these wares contributed to solving hardware problems. The case of organic agriculture demonstrates that engagement of societal actors early in the process of innovation reduces the challenges and risks an invention meets on the way to wider adoption.

2.3.1.2. GM crops

The invention of GMO has had a very broad impact on all aspects of life sciences. The agricultural green biotechnology with GM plants and animals is still under critical discussion. Because of that, regulation for the release of GM plant varieties is highly complicated in Europe. Negotiations on GMO-regulations and the import of the first GM soybean (resistant to herbicide glyfosate) was the starting point for open disagreement between societal parties involved in the debate.

The combined invention of gene cloning and genetic modification of micro-organisms, plants, animals and for human health was originally science driven but has adopted many intersectional aspects. GM crop production has become the main example of limited application of an invention in important global markets due to its lack of acceptance within society. The history of this invention distinctly differs from organic agriculture, yet has one interesting similarity. One of the main differences is the starting point. In organic agriculture the starting point was focused on a philosophical idea on how to go forward with more sustainable agricultural processes. The starting point of the GMO invention became more focused on the approach and due to misalignment with societal drivers focused almost exclusively on rules concerning the risks of the technology rather than the benefits and solution of sustainability issues (Anonymous 1998).

An important other difference is the outcome of discussions about GMO compared to organic agriculture with the involved stakeholders and the prevalence of safety issues. Interestingly, these discussions on safety regulation development involved the same type of actors as those involved in the discussion on organic regulations: government, science, branch organizations and non-governmental organizations (NGOs). What caused a different outcome of discussion? First of all, the GM invention was a science driven invention, rather than society driven. At the start, the actors that were involved in the discussion had different backgrounds and starting points about the desirability and need for the application of GMO and, therefore, were not involved in the underlying philosophy of the invention. Also, opposite to the organic agriculture debate, large industries were involved in the discussion from the early days, introducing a different power balance. This all caused disagreement and lack of harmony in the discussion, eventually resulting in a societal drive towards detailed description and application of rules in the different fields of GMO application, especially in the European Union.

The outcome today is that the invention of GM-crop production is very successful outside Europe. In most cases only multinational companies are able to develop GM-varieties, due to the strict regulations. These companies can spend enough money to fulfill the highly demanding requirements that are needed to allow release of a GM variety onto the market. This makes GM-varieties very expensive. The estimated cost of getting a single GM-variety to the market is 6 to 7 million euro (Schenkelaars 2008). This is much higher than conventionally bred varieties, where expensive safety regulations are not playing a role, and makes

GM technology mainly applicable to big field crops, such as maize, cotton and soybean. The protests of the NGOs against GM crops did improve the position of multinationals in this field and weaken that of smaller, private breeding companies. In 2009, worldwide, over 140 million hectares of GM crop plants were grown, mainly consisting of maize, soybean and cotton.

2.3.2. Directional inventions

Directional inventions originate within a specific area of expertise (Johansson 2004). Directional ideas are operating within a specific field and are mostly based on specific expertise within that field, which is extended. A field relates to a discipline, culture or domain such as music, (molecular) biology or software industry in which one can specialize in a mono-disciplinary way. Directional inventions can, for example, help to simplify regulations based on reconstructed and/or extended knowledge.

2.3.2.1. Cisgenesis

In the field of genetic modification, cisgenesis is a recent development and considered as a directional invention. Central to the philosophy of cisgenesis is that it takes the grounds of opponents of GM crops as a starting point, and uses the opponents' suggestions where applicable.[4] In cisgenesis, natural genes (cisgenes) from either the plant itself or from a crossable species are used. Hence these genes are not new and belong to the existing breeders' gene pool with a history of safe use. Based on NGOs' and societal opponents' arguments against GM crops, cisgenesis applies marker free transformation with only cisgenes (Schouten et al. 2006a, 2006b).

Transgenes are genes coding for traits which are (partly) based on genetic information found in non-crossable species like bacteria and viruses. They are thus alien to the plant they are imported into. Therefore, transgenes are perceived as being 'unnatural'. For many, the use of transgenes is not acceptable rather than the technique of transformation.

Transformation of plants with transgenes or cisgenes is done by the bacterium *Agrobacterium tumefaciens*. In nature this soil bacterium transfers a few own genes, present on transfer-DNA (TDNA), into individual cells of the plant to induce a crown gall with production of amino acids for pathogens' own use. Normally these genes are not inserted into cells involved in sexual reproduction. However, during evolution of 15 different tobacco species up to 4 genes of this

---

[4] It is revealing that the idea of cisgenesis was invented through discussions within a Dutch Christian political party (Jochemsen 2000), while in general genetic modification of organisms is rejected by religious movements.

TDNA were introduced into the plants' sexual reproduction system (Intrieri and Buiatti 2001).

Initially, only transgenes were involved in the discussion on GMO regulations. For plant breeding, transgenes belong to a new gene pool, so that a demand for additional safety tests is logical. But, although society driven, cisgenesis is still being discussed in the societal debate. Based on the integration of societal concerns, Schouten et al. (2006a, 2006b) suggest that cisgenesis might be exempted from the GM regulation following the earlier rationale for induced mutations and protoplast fusion between crossable species.[5] Intermediate steps for application could consist of derogation for specific trait-crop combinations (Jacobsen and Schouten 2007, 2009).

By attempting to overcome the most important complaints of society against GM-plants with transgenes, the new technique of cisgenesis showed increased acceptance of the public (Duchateau 2008; Sameer et al. 2009).

It is clear that this invention is the result of public debate which requires an adequate balance between hardware, software and orgware. It is a directional invention with respect of type of genes used that potentially influences existing regulation. Both organic agriculture and cisgenesis supporters have realized a dynamic decision-making in the higher echelons of governments.[6] Regulators are continuously amending and adapting rules and regulations to cover new developments and insights in science, technology, practice and society (Yanarella 1975).

2.3.3. Open innovation

In agricultural life sciences, open innovation was the standard as described in earlier parts of this chapter. New scientific insights were published and freely available for everybody. However, in plant breeding, seed business developed a sui generis system called PBR to protect varieties but to keep them freely available as breeding parent for development of new varieties. The Bayh-Dole Act (1980) stimulated the introduction of patent rights which changed the open situation in agricultural life sciences. The recent experiences in modern plant breeding showed potential negative side effects of PR in this field. We believe that open innovation has the future, especially when it has to be related to sustainability because open innovation systems have the potential to lead to a quick succession of incremental innovations, which is needed in new fields. In addition, in certain fields communities of practice are frequently needed for solving problems in order to go quickly into the right direction. We are describing a few recent examples below.

---

[5] Annex 1b of the EU Directive 2001/18/EC.

[6] Directive 2001/18/EC for GMOs and Council Regulation (EC) no. 834/2007 for organic production.

2.3.3.1. Care farming

Care farming is an upcoming activity in the agricultural sector (Hassink and Dijk 2006). Care farming is an unusual combination of disciplines like (organic) agriculture, psychiatry and transition sciences (see for more information on care farming action experiment Green Care in Appendix II). Care farming is a win-win initiative for clients, farmers and the community as a whole. The clients range from people with a mental disability to youth with behavioral problems, addicts and psychiatrically patients.

For the invention of care farming to become a fully fledged innovation, it has to solve a unique problem in agriculture as its clients, psychiatric patients, are not traditionally within the remit of agriculture. Support is therefore needed from the medical world. While both care farmers and their clients are very positive about the results (Elings and Hassink 2008), the relevance of combining the care of people with mental disorders with farming activities is still unclear. To make care farming even more attractive for, among others, health care insurers, interdisciplinary scientific research is highly needed to provide a scientific evidence base of the potential health effects of combining agriculture and care for psychiatric patients and specifically to gain insights into the health promoting elements of agriculture (Leerink 2009).

The lack of patents or other IPR systems on the basic invention makes care farming an open innovation with many characteristics of an intersectional invention. Care farms are a good example where an important innovation may be achieved without the need to develop too much specific (disciplinary) expertise.

2.3.3.2. Energy producing greenhouses

Currently, in the Netherlands, the greenhouse industry contributes to approximately 10% of the total natural gas consumption and up to 10% of the electricity use. However, greenhouses can reduce their fossil energy use considerably. The invention of an energy producing greenhouse came up in the early years of the 21st century. The energy producing greenhouse is a greenhouse that needs a small fraction of the energy needed by the most common greenhouses today and is even expected to be net producer of energy (Anonymous 2009b). Some recently developed energy efficient greenhouse concepts are based on durable energy sources such as solar energy, wind energy or geothermal energy: the solar greenhouse (Bot et al. 2005), the closed greenhouse (Opdam et al. 2005), the energy producing greenhouse (Bakker et al. 2006) and the Sunergy greenhouse (De Zwart 2009). A current example is a greenhouse system which converts natural energy sources such as solar energy into high-value energy such as electricity. Sonneveld et al. (2007) designed a system with a parabolic NIR (Near InfraRed) reflecting greenhouse cover. This cover reflects and focuses the NIR radiation on a specific PV (photo voltaic) cell to generate electricity (Electricity producing greenhouse).

One of the bottlenecks in applying this invention of an energy producing greenhouse concerns the physiological aspects of the plant. This new greenhouse has a completely different type of climate than conventional greenhouses. Air humidity has changed and new ways are needed to control crop performance. A new way of growing plants, called 'conditioned growth', manipulates the temperature of the different plant organs. This invention requires an iterative process of greenhouse technology and crop growth physiology (Bezemer et al. 2009). The action experiment SynErgie (see Appendix II) experimented with the application of an energy producing greenhouse. Hardware and software elements are of major importance to realize this conditioned growth. Free exchange of knowledge and entrepreneurship in this case is of crucial importance as it is the grower who needs to use this invention.

Modern greenhouse culture is changing more and more into sustainable vegetable production, serving great numbers of consumers with healthy food. In the near future, greenhouse horticulture will be more than just a supplier of high quality fruit, vegetables or flowers. They will also supply their excess solar energy to third parties in the form of heat or electricity.

Where the connection between care and agriculture was an open innovation between different sectors, connections between scientific fields can also be the result of directional innovations which improve a product in predictable steps along a well defined dimension (Johansson 2004). The inventive intersectional idea of the energy producing greenhouse is a nice representative of this, as it requires in specific fields directional improvements on greenhouse design, energy validation, and plant growth models.

Another major development is the increased influence of society on new developments, such as illustrated by the concept of energy producing glasshouses and the need of care farming. It illustrates that open innovation today in agriculture is still very important and often a guaranty for determining the right direction of new developments for sustainable agriculture as quickly as possible.

2.4. Conclusion

We started this chapter with two questions. What aspects are of importance to turn inventions into successful innovations and whether it is possible to select inventions, including an understanding of their limiting conditions, so that they can improve sustainability? We have addressed some past inventions in the agricultural sector and how they were applied into innovations by the linear process of fundamental, strategic and applied research. Sustainable development of the agricultural sector requires system innovation. Recently, inventions have been followed more frequently by non-linear innovation processes, they will play a crucial role to realize such system innovation, as is shown in this chapter. We have seen that inventions can be science driven, such as the invention of dwarf mutants, the hybrid growth concept and plant biotechnology. Inventions can also be society driven, if societal demands are the motivation for a change, which in turn requires inventions. Examples are energy producing greenhouses, care farming, and organic agriculture. The need for a more sustainable development of the agricultural sector is a societal demand for new inventions that stimulate the needed transition of the agro innovation system. In order to be more frequently successful such society driven inventions should take several aspects in consideration. Continued interaction with society is needed to direct these innovations towards acceptable designs. However, if relevant stakeholders in society, such as the immediate surrounding, are not adequately informed and/or involved from the beginning, obstruction against the final implementation of even society driven inventions and innovations may be possible.

In general we state that to realize inventions for a more sustainable agriculture, and turn these inventions into successful innovations, several aspects are of importance:

- The process of implementation of inventions into innovations will become increasingly non-linear. From the beginning of an innovation balancing of hardware, software and orgware elements of an invention through entrepreneurship are needed. This requires trans-disciplinary, rather than basic research.
- Intellectual property rights in agricultural life sciences can obstruct the necessary implementation and dispersal of an invention into an innovation, and therewith hinder the way towards a more sustainable agriculture. Sustainable development requires rapid sharing of new techniques and (re)combining existing inventions. The increasing use of IPRs in the agricultural sector can stimulate but also frustrate the innovation process. Non-voluntary licensing of patented inventions might be a theoretical but not practical solution to such problems from the beginning.
- When confronted with the challenge to find new solutions for sustainability and one doesn't know where one is looking for, organizing intersections between different disciplines can come up with new inventions that combine expertise from different fields. This approach could later be made more specific by organizing directional innovations.
- If the solution can be found in a specific field, such as a specification of an existing technique, organizing directional inventions enables a specific search for an invention.
- In case of highly complex problems of which the solution is unknown, new inventions can come up by using the collective intelligence of several actors with either specific or different disciplines in an open innovation context. The development of an energy producing greenhouse is an important example of such an approach.

Acknowledgments

The authors were inspired by Dr Henk van Latesteijn and Karin Andeweg from TransForum. We acknowledge textual suggestions in their field by Dr Anja Dielemens (Wageningen University and Research) and Dr Marjolein Zweekhorst (Vrije Universiteit Amsterdam).

2.5. References

Anonymous (1998). Regeling genetische gemodificeerde organismen en richtlijnen van de COGEM bij deze regeling. Commissie Genetische Modificatie.

Anonymous (2003) Invention and innovation for sustainable development. Report of a workshop sponsored by the Lemelson_MIT Program and LEAD International, London, November 2003, 1–42.

Anonymous (2009a). UPOV, what is it, what does it? UPOV publication 437 June 2009. http://www.upov.int/index_en.html.

Anonymous (2009b). Innovation and Demo Centre: Greenhouse as energy source. Brochure pp 7. www.energiek2020.nu.

Bakker, J.C., De Zwart, H.F., and Campen, J.B. (2006). Greenhouse cooling and heat recovery using fine wire heat exchangers in a closed pot plant greenhouse: design of an energy producing greenhouse. *Acta Hort*, *719*, 263–270.

Bevan, M. (1984). Binary Agrobacterium vectors for plant transformation. *Nucleic Acid Research, 12*, 8711–8721. http://nar.oxfordjournals.org/misc/terms.shtml

Bezemer, J., Bontsema, J., and Weel, P.A. van, (2009). Vocht afvoeren zonder minimumbuis: Met verticaal ventileren gelijkmatiger klimaat en lager energieverbruik. *Interview Onder Glas, 6*(2), 8–9.

Borlaugh, N.E. (2000a). The green revolution revisited and the road ahead. Special 30th Anniversary Lecture, The Norwegian Nobel Institute, Oslo, September 8, 2000, 1–22.

Borlaugh, N.M. (2000b). Ending world hunger. The promise of biotechnology and the threat of antiscience zealotry. *Plant Physiol, 124*, 487–490.

Bot, G., Van de Braak, N., Challa, H., Hemming, S., Rieswijk, T., Van Straten, G., and Verlodt, J. (2005). The solar greenhouse: state of the art in energy saving and sustainable energy supply. *Acta Hort, 691*, 501–508.

Branscomb, L.M. and Auerswald, P.E. (2002). Between invention and innovation: An analysis of funding for early-stage technology development. Economic Assessment Office. Advanced Technology Program. National Institute of Standards and Technology, 1 138. http//:www. Atp.nist.gov.

Cohen, S.N. and Boyer, H.W. (1980). Process for producing biologically functional molecular chimeras US PATENT: 4237224 SERIES CODE: 6; ISSUE DATE: 19801202.

Crow, J.F. (1998). 90 Years ago: The beginning of hybrid maize. *Genetics, 148*, 923–928.

De Zwart, H.F. (2009). The sunergy greenhouse – one year of measurements in a next generation greenhouse. http://edepot.wur.nl/14135.

Druckner, P.F. (1993). Innovation and entrepeneurship. *First Harper Business Edition*, 1–269. ISBN 0-88730-618-7.

Dubois, L.M.A. (2001). Intellectual property rights on plants with special reference to vegetatively propagated ornamentals. *Acta Hort, 552*, 207–214.

Duchateau, I. (2008). Public reactions toward cisgenic apples. Report by NewCom Research & Consultancy.

Elings, M. and Hassink, J. (2008). Green Care Farms, A safe community between illness or addiction and the wider society. *Int J Ther Commun, 29*, 310–321.

Hassink, J. and Dijk van, M. (2006). Farming for health: Green-care farming across Europe and the United States. Springer Dordrecht, Netherlands.

Intrieri, M.C. and Buiatti, M. (2001). The horizontal transfer of Agrobacterium rhizogenes genes and the evolution of the genus Nicotiana. *Mol Phylo Evol, 20*, 100–110.

Jacobsen, E. and Schouten, H.J. (2007). Cisgenesis strongly improves introgression breeding and induced translocation breeding of plants. *Trends Biotechnol, 25*, 219–223.

Jacobsen, E. and Schouten, H.J. (2009).Cisgenesis: An important sub-invention for traditional plant breeding companies. Euphytica. DO10.1007/s10681-009-0037-y.

Jochemsen, J. (2000). Toetsen en begrenzen: een ethische en politieke beoordeling van de moderne biotechnologie. Uitg. van de wetenschappelijke instituten van GPV en RPF. ISBN 9072016327, pp. 262.

Johansson, F. (2004). *The Medici effect: What elephants and epidemics can teach us about innovation.* Boston, MA: Harvard Business Press.

Leerink, B. (2009). Dag van de ZorgLandbouw. Accessed on May 1st, 2009 at http://zorglandbouw.nl/programma/index.php.

Lelyveld van, P. (2009). Veredeling en manipulatie: tumult rond biotechnologie in industrie, wetenschap en politiek. www.uitgeverijbalans.nl, ISBN978-94-600-30130.

Leeuwis, C., Smits, R., Grin, J., Klerkx, L.W.A., van Mierlo, B.C. and Kuipers, A. (2006). Equivocations on the post privatization dynamics in agricultural innovation systems. In: *The design of an innovation enhancing environment. Working papers no. 4* (pp. 3–58). Zoetermeer: TransForum Agro & Groen.

MacGill, I. and Outhred, H. (2004). CAN Europe Workshop. The policy and environmental implicatios of CO_2 capture and storage, hydrogen and fuel cell technologies. http://wpw-ergo.ee.unsw.edu.au/can%20europe%20ccs%workshop%202004%20-%ifm.pdf.

Malthus, T.R. (1798). *An essay on the principle of population.* Oxford World's Classics.

Marton, L.C., Zell, S. and Szieberth, D. (2003). Past results and future prospects of hybrid maize breeding in Hungary. Bericht über die 54. Tagung der Vereinigung van Pflanzenzüchter und Saatgutkaufleute Őterreichs BAL Gumpenstein, 25–27 November 2003, pp. 1–5.

Monson, R.K. and Moore D. B. (1989). On the significance of C3-C4 intermediate photosynthesis to the evolution of C4 photosynthesis. *Plant, Cell Environ, 12*, 689–699.

Norman, A.L. (2005). Technical Foundation of the Modern Economy, Subchapter: Discovery, Invention and Innovation, Paragraph: Why the invention rate exceeds the innovation rate. In: *Information Society Notes*; Internet version.

Opdam, J.J.G., Schoonderbeek, G.G., Heller, E.M.B. and De Gelder, A. (2005). Closed greenhouse: a starting point for sustainable entrepreneurship in horticulture. *Acta Hort, 691*, 517–524.

Reichman, J.H. and Hasenzahl, C.J.D. (2003). Non-voluntary licensing of patented inventions. ICTS-UNCTAD. Project on IPSs and sustainable development. Issue Paper No. 5.

Rimmer, M. (2003). Franklin Barley: Patent Law and Plant Breeders' Rights. Murdoch University Electronic Journal of Law 10 No. 4, Available at SSRN: http://ssrn.com/abstract=603228.

Salamini, F. (2003). Hormones and the green revolution. *Science, 302*, 71–72.

Sameer, J., Soriano, J.M., Schaart, J., Broggini, G., Szankowski, I., Jacobsen, E., Krens, F.A. and Schouten, H.J. (2009). Approaches for development of cisgenic apple. *Transgenic Plants J, 3*, 40–46.

Schenkelaars, P. (2008). Dossierkosten markttoelating genetisch gemodificeerde gewassen in de Verenigde Staten en Europa. COGEM.

Schouten, H.J., Krens, F.A. and Jacobsen, E. (2006a). Do cisgenic plants warrant less stringent oversight? *Nat Biotech, 24*, 753.

Schouten, H.J., Krens, F.A. and Jacobsen, E. (2006b). Cisgenic plants are similar to traditionally bred plants. *EMBO Rep, 7*, 1–4.

Smits, R. (2002). Innovation studies in the 21st century: questions from a user's perspective. *Tech Forecasting Social Change, 69*, 861–883.

Smits, R.E.H.M. and Boon, W.P.C. (2008). The role of users in innovation in the pharmaceutical industry. *Drugs Discov Today, 13*, 353–359.

Sonneveld, P.J., Swinkels, G.L.A.M., Bot, G.P.A. and Flamand, G. (2007). Conversion of NIR-radiation to electric power in a solar greenhouse. Thermovoltaic generation of electricity: TPV7: Seventh World Conference on Thermophotovoltaic Generation of Electricity. AIP Conference proceedings, 890, 317–326.

Ten Napel J., Bianchi, F. and Bestman, M. (2006). Utilising intrinsic robustness in agricultural production systems. In: *Inventions for a sustainable development of agriculture. Working papers no. 1* (pp. 32–54). Zoetermeer: TransForum Agro & Groen.

Termeer, C.J.A.M. (2009). Barriers to new modes of horizontal governance. *Public Management Rev, 11*, 299–316.

Yanarella, E.J. (1975). "Reconstructed logic" & "logic-in-use" in decision-making analysis. *Polity, 8*, 156–172.

Zilberman, D., Yarkin, C. and Heiman, A. (2001). Privatization and innovation in agricultural biotechnology. Lecture Notes, University of California.

Chapter 3. Organizing innovations and transitions

Toward a more sustainable fit between innovation strategies and the institutional environment

Hans Mommaas[1] **and Rik Eweg**[2]

[1]J.t.mommaas@uvt.nl; TransForum and Telos, Tilburg University, P.O. Box 90153, 5000 L Tilburg, The Netherlands

[2]Eweg@TransForum.nl; TransForum, P.O. Box 80, 2700 AB Zoetermeer, The Netherlands

Abstract This chapter concentrates on the interaction between developmental dynamics in the different action experiments, aimed at exploring new more sustainable agricultural business models, and changes in the wider institutional environment. Which (mis)fits can be identified? What has to be done to enable agro innovation to more effectively be managed in a rapidly globalizing *and* urbanizing environment? Based on a loose theoretical framework, consisting of notions of the network society, transition theory and urban regime theory, we come to the conclusion that we need a new conceptual/strategic story line, with the developmental coalitions related, enabling a productive fit between a 'transitionalizing' landscape of agriculture on one side and a highly urbanizing metropolitan environment on the other.

3.1. Introduction: Context of the research

This chapter focuses on organizing innovations and transitions in the agricultural sector. The main question addressed is how to facilitate effective transition processes enabling agro production systems to adapt to new and more sustainable societal needs and ambitions. The insights presented are based on a number of action experiments that have been conducted in the past 5 years (see Appendix I for more information). In discussing the first outcomes, we especially focus on misfits between the innovational dynamics unleashed and the wider institutional environment, thus far emerging from the action experiences. In doing so, we have to take into account three characteristics that are specific for the Dutch context of the action experiments:

H.C. van Latesteijn and K. Andeweg (eds.), *The TransForum Model: Transforming Agro Innovation Toward Sustainable Development*, DOI 10.1007/978-90-481-9781-1_3,

First, the action experiments took place in an agro production system which, over the course of time, has developed into one of the most efficient ones globally, both in economic and in ecological terms. With the United States and France, the Netherlands are amongst the top three important exporters of agricultural produce globally (USDA 2005). Second, as such, the action experiments take place in parts of the Rhine-Scheldt Delta area, a highly fertile environment, which however at the same time is situated in one of the most densely populated areas globally (UN 2006). Third, partly because of this situation, the Dutch agro production system is recently confronted with some fundamental problems. These not only have to do with the high intensity of the production system as such and its admissibility to developmental risks related (c.f. notions of the risk society, see Beck 1986). In addition, the problems have to do with the urbanized context in which it operates. This involves more than pricing pressures related to the scarcity of space. With urban economies involved in an increasing worldwide competition for talent, investment opportunities and 'quality of life' conditions, the surrounding countryside is more and more evaluated as not just a production environment, but increasingly also as an environment of consumption, recreation and experience (Marsden 1998; Mommaas 2003). This is confronting the agricultural sector with new demands, both with regard to landscape qualities, and with regard to issues of food-quality and animal welfare. As a result, the countryside is rapidly being integrated in a new socio-spatial reality, beyond a former rural-urban divide, facing agriculture with the challenge to strike a new balance between its international role as an efficient techno-economic production system and its regional role as a resource of a quality of life (e.g. Andersson et al. 2009).

Against this background, the specific challenges for agro innovations situated in a highly urbanized but at the same time highly fertile delta area become clear. How to strike a new, more sustainable balance between the role of the agro system as an efficient global food production system, and its role as an important caretaker of a highly urbanized and thus highly valued landscape (e.g. Veldkamp et al. 2008)?

In the following, we first want to go into some of the more basic theoretical threads which have underlined and motivated the research. We hope to keep this brief, but on the other hand we need this to make the reader clear why we have chosen the approach concerned. Next, and in line with our theoretical approach, we want to go into experiences gained with regard to innovation strategies primarily embedded in what we will typify as the 'space of flows' of agricultural production. Which obstinate problems can be identified? Following that, we will pay attention to more place-embedded transition strategies. Which factors are hindering transition processes here? From the outset we expected these different spaces of innovation to reflect some fundamental tensions. Last, we would like to integrate both fields of experiences and extract major innovation obstacles. What is currently hindering the transformation of an agricultural system – such as the Dutch one situated in a highly productive but at the same time highly urbanized environment – toward a more sustainable, 21st century production system? Which decisions have to be taken to surpass the lock-ins, and mobilize more sustainable developmental opportunities?

3.2. Theoretical building blocks

In this chapter a lose theoretical framework is applied to enable a comprehensive analysis of research brought together from different disciplines and theoretical perspectives. In the course of time, this framework facilitated the wider knowledge development without becoming something of a theoretical straightjacket in bridging the different knowledge levels and links to the action experiments.

Three theoretical building blocks were of pre-eminent importance. First, we used insights from Castells' notion of the global network society (Castells 1996, 2009) to come to grips with some of the fundamental dynamics and dialectics of the contemporary economic condition, operating within the contexts of a highly globalized and information and communication rich environment. Basic here was Castells' notion of the increasing tension between the 'space of flows' and the 'space of places'. Within the contemporary global society, networks have come to constitute the fundamental pattern of life. Of course, as Castells himself aptly notices, networks are not specific to 21st century society. Also in the past there have been important historical episodes in which the changing power of networks fundamentally changed socio-spatial conditions, also within the realm of agriculture. However, in a certain sense, in the pre-electronic age, these changes have always been limited in size and scale, thus in the end favoring vertical-hierarchical command and control systems. According to Castells, it is only with the advent of post 1970s information and communication technologies, developing in close interaction with economic and societal opportunities, that the power of networks has been unleashed to the point where they have become the most efficient organizational form (Castells 2009: 30).

We don't necessarily have to agree with Castells' historical outlook (it is questionable whether his somewhat teleological representation will in the end do justice to the situated character of human history) to appreciate the insights this produces with regard to the contemporary economic condition. Basic are (a) the organizational advantage of networks, and the innovation and valuation logic related, over more vertical-hierarchical forms of organization, and (b) the increasing tension this produces between social forms of time and space dominated by the logic of networks, and social forms of time and space dominated by the logic of contiguity. While in the space of places, meaning, function and locality are closely interrelated, in the space of flows localities owe their function and meaning from their nodal role in the specific networks in which they operate (Castells 2009: 34). Still struggling with the consequences of the first debt-crisis of the beginning of the 21st century, it becomes abundantly clear how the current unleashing of network dynamics is posing some fundamental problems in managing the relationship between markets and communities, both in the wider sense of nation-states, as in the more everyday sense of regional localities.

A second theoretical building block was taken from innovation/transition theory. We were interested in representations of how innovation and/or transition trajectories 'work' and what stimulates or hinders their success or failure from a broader, integrated 'micro-macro' perspective. Here we took recourse to the multilevel viewpoint on innovation/transition trajectories such as developed by Geels, Schot and colleagues (see Geels et al. 2004; Geels and Schot 2007; Schot and Geels 2008). In trying to comprehend the dynamics of innovation and transition processes, especially with regard to their societal successfulness, they heuristically differentiate between three levels of change: niche, regime and landscape. In generic terms, the more niche innovations are able to take into account emerging landscape developments, the more they are successful in overthrowing established regimes, in favor of new ones, thus producing a true transition. In a first instance these cycles of innovation/transition were regarded as something of a revolutionary nature; old regimes have to be overthrown to enable new developmental directions. However, especially with regard to dynamics in the agricultural system, more evolutionary trajectories surfaced, with regimes being more sensitive to change.

Despite the somewhat awkward spatial metaphor implied (both below-above and here-there metaphors obstructs a clear view on the actual workings of the integration processes involved), the innovation/transition approach is relevant because of the way it addresses the interaction between different scales of social organization, thus balancing processual/actor and institutional/structural aspects. This is important in order to prevent either actor-centric of structure-centric forms of reductionism in thinking through the dynamics of innovation processes, instead focusing on how, from a practical point of view, scales of social/system integration constantly interact, either in a positive or in a negative sense (c.f. Klerkx et al. 2010).

Third, central in this multi-scale perspective is the notion of 'regimes'. These are becoming increasingly important in a social environment dominated by the power of networks. Contrary to vertical-hierarchical forms of organization, regimes represent institutionalized forms of horizontal collaboration. As a theoretical concept, the notion partly stems from the field of international relations theory (e.g. Ruggie 1975; Krasner 1983). Here, it represents the idea of groups of nations building systemic relationships with one another around a common agenda. In this research we use insights developed in the field or urban/spatial planning/geography. Urban Regime Theory is about stable, informal forms of public-private collaboration (see e.g. Stone 1989, 1993; Stoker and Mossberger 1994; Mossberger and Stoker 2001). The theory stresses the importance of public-private bridging mechanisms in order to 'get things done'. Urban (and regional) regimes thus represent collaborative or network based ways of doing things, bridging social organizations, without these collaborations being taken up in a new formal-hierarchical organization. Regimes thus have a history and solidity, based on proven trust and on the fact that partners need each other in order to pursuit their interests. They bring together divers resources, such as knowledge, money, licensing, around a shared public agenda – mostly rather a set of shared values, viewpoints, affinities or interests than a formalized agenda.

In Urban Regime Theory, regimes represent a 'good' thing. They organize a capacity to act, otherwise absent. In Transition Theory, on the other hand, regimes represent a 'bad' thing, something hindering innovation and development, in need of being overthrown. Given different circumstances, both viewpoints may be true, and thus both schools tend to overemphasize just one particular moment in a broader field of developmental possibilities. Regimes are both enabling and constraining, depending on their moment of development, changing circumstances and the different interests/perspectives involved. Here, we need the notion to understand how in the various sections and moments of agricultural development public-private production regimes have come into existence, favoring specific developmental models and dynamics. How are these either enhancing or resisting current needs for change, given changing societal (e.g. 'landscape') circumstances. In what sense are 'niche' developments accommodating 'regime' changes or being hindered by them?

Together, these theoretical building blocks delivered the broader rationale, underlying this research. As such, they have also informed the structure of this chapter. In line with Castells' analysis, we will first of all focus on innovation-attempts predominantly to be situated in the logic of the space of flows. Next, we will focus on action experiments primarily developed out of contexts of spaces of place. Subsequently, we will try and bring the two fields of experiences together, thus tearing out the fundamental 'niche-regime' or 'micro-macro' blockages identified.

3.3. Coming in from the space of flows

Looking at agriculture from the perspective of the space of flows primarily implies looking at how agricultural production/distribution is organized in organizational and/or informational networks, spanning the space from 'earth' to 'mouth'. In general terms, sustainable development here implies looking at ways in which systems of exchange/interaction within and between production/distribution chains can be made more efficient/resilient in terms of 3P values (people, planet and profit). This might imply looking at e.g.:

- breeding/developing products better adapted to (changes in) the landscape/climate/economy at hand, thus for them to need less labor power, energy, pesticides;
- increase the eco-efficiency of existing productions procedures in terms of uses of substances, energy, labor, investment costs, animal welfare;

- re-arranging existing production/distribution chains, both from a vertical and from a horizontal perspective, e.g. by creating cross-sectoral flows of (rest)substance, by reducing production-consumption cycles, by re-arranging the distribution of power/wealth along chains;
- strengthening the role of consumers as potential change agents, e.g. by mobilizing relevant market information, or by more directly involving consumers as co-creators of more sustainable markets.

3.3.1. Innovation as reconfiguration

Overlooking the thirty action experiments, some conclusions are evident. A first important and overarching insight is that while transitions and system changes are usually thought to come about through breakthroughs of technological discontinuities, a closer look at current and past developments suggest more gradual, stepwise 'reconfigurations' as an important transition trajectory (Geels et al. 2008). In this kind of trajectory, innovations are adopted in the existing system, and gradually reconfigure the basic architecture of the existing regime. New combinations of 'old' and 'new' elements gradually change the system in a stepwise fashion. It expresses an awareness in the sector of the need to change the current developmental regime, to make it more adaptable to changing needs and the accompanying value system implied. These type of 'reconfiguration' trajectories deviate from breakthrough transitions in three aspects:

- the process is not driven by *one* major, radical innovation, but by *multiple* (component) innovations;
- these innovations do not compete with the existing system, but are incorporated as add-ons or component replacements within existing supply chains; transitions then do not consist of fights between 'old' and 'new' technologies, but are more gradual processes in which new combinations of 'old' and 'new' increasingly change the system's architecture in a stepwise fashion;
- present actors are not swept away by new entrants (as in 'waves of creative destruction'), but survive the process; current actors enact the reconfiguration of the system architecture; the *development* of the innovations, however, is often done by other (outside) actors. Hence, the transfer of knowledge and innovations to sitting actors is an important aspect of reconfiguration transitions (ibid.).

As the research points out, in bridging the old and new situation, it is important that different forms and expressions of boundary spanning or *anchorage* (technological, social and institutional) are closely intertwined and logically connected.[1] Innovation processes in one way or another have to create alignments within and between the fields of technological development, the network of involved stake-

---

[1] Chapter 2 elaborates on the hardware, software and orgware elements of an invention.

holders and the wider institutional landscape. Earlier episodes of anchorage create the conditions for later forms and there is the constant challenge to work on the fit between them. Here, there is a crucial initiating role for *hybrid forums/actors* which operate at the intersection between niches and regimes/landscapes. Besides, research points at the importance of increasing normative pressures (to create a sense of urgency and make actors line up in the same direction), and of alignments with or spill-overs to economic, regulatory, socio-cultural and technological niche developments (ibid.).

Completely in line with the overall literature on innovation/transition management (see e.g. Grin et al. 2010), the research expresses a warning that one should not be overtly optimistic about the scope for planning and controlling innovation processes. However, this does not render deliberate interventions and projects meaningless, on the contrary. The innovation trajectories implied involve and weave together a wide variety of networks (including hybrid forums) and developments that are somehow part of (pilot) projects, programmes and interventions. This is in line with both evolutionary understandings of innovation processes (e.g. Hoogma et al. 2002) and approaches which build on theories about complex dynamic systems (Prigogine and Stengers 1984; Loorbach 2007; Leeuwis and Aarts 2008). In a context in which it becomes more and more difficult to predict the future in a linear kind of fashion, innovation programmes organize the necessary trial and error spaces. They work as spaces of experimentation and sensibilization, bringing stakeholders together in new configurations around a common awareness, exploring possible directions for change. At the same time, they organize the kind of resources (both in terms of new knowledge, new contacts, venture capital, licensing space) which enable entrepreneurs to explore yet uncertain opportunities, and learn from them.

3.3.2. Between 'business as usual' and 'market failures'

Whereas larger companies are fully capable of organizing these evolutionary innovation/transition spaces for themselves (e.g. in forms of corporate venturing, mostly in collaboration with strategic partners), on the level of broader sectors and/or clusters, at interstitials between sectors, and at the level of small and medium sized enterprises, there is ample room for public-private programmes, creating the kind of space which invites knowledge brokers, entrepreneurs (both large and small and from different regimes/sectors/networks), and venture capital to come together and explore yet unknown probabilities and risks. These public-private innovation systems are considered legitimate in the face of market or system failures hindering more independent innovation processes. In fact, their primary goal could and should be to identify these market failures and come up with solutions to deal with them in a more structural sense.

There are lots of critical issues involved in organizing/managing such adaptive innovation/transition systems. Research reveals issues of type, form, moment and number of stakeholder involvement; effective levels of complexity; the formal/ informal organization and management of network dynamics; the organization of trust and the related balancing of moments of exploration and exploitation; the facilitation of adequate learning environments, building-in productive monitoring systems, but also leaving ample room for surprise; the enabling/inhibiting role of dominant players in relation to newcomers; balancing local and global relationships; enabling/inhibiting forms of finance and investment, et cetera. In general terms, these issues have become part of what might already be regarded as something of a 'normal' innovation policy and/or transition management landscape, together with the models and insights related (see e.g. Nooteboom and Stam 2008; Poppe et al. 2009; Grin et al. 2010).

At the same time, organizing and researching such an innovation landscape has indeed raised a few issues which go beyond 'business as usual'. We regard them as the primary source of the market failures currently hindering innovative developments in agribusiness, and thus in need of firm institutional action.

A first set concerns the *access to consumer markets*. There is a growing awareness of the fact that, in the end, the access to consumers is a fundamental preconditions for a more sustainable development/innovation of food supply chains. Of course there is still ample space for an increase in eco-efficiency within the existing food production system, completely in line with a dominant economics of scale. We will come back to this later. However, value creation and risk taking amongst producers demands financial and communicative space, enabling experimentation. Important here is not only the value distribution amongst the production – wholesale – retail – consumption chain, important is also the ability for producers to position their products, together with the storylines related, more proactively in consumer markets.

Over a broad spectrum of products, food producers are complaining about the fact that they don't have the opportunity to adequately confront consumers with alternative products and the storylines related. As a consequence there is a reluctance to take added risks. Here we stumble upon a stubborn problem which, already for some time, is playing its role in the field of agro food markets. The access to consumer markets seems to be faced with serious lock-ins. At the same time, there is a fierce discussion about what is producing this situation and who is to blame (e.g. Bunte et al. 2009; NMA 2009). Is this primarily an issue of consumers not being interested in paying a higher price for more sustainably produced foodstuffs, or is this about large scale retailers keeping each other and the consumer market hostage in a battle for ever cheaper foodstuffs, thus hindering alternative value capturing processes? What is the role of service providers and the wholesale sector, together with the retailers possibly keeping larger segments of production-distribution regime hostage in a collective economics of scale? Besides, there is still also the possibility that producers are too strongly focussed on supply dynamics, not taking into account emergent demand characteristics, a classic situation of dogs barking up the wrong tree. The overall challenge here is to more critically look at the institutional dynamics in and around food markets, preferably in comparison with international experiences, and in close interaction with consumer practices, in order to include those dynamics more reflexively in innovation trajectories, exploring new value propositions in consumer markets.

One of the action experiments, an experiment designed to provide pre-competitive information on more sustainably produced foodstuffs, suggests that this is more than merely an information issue (see Appendix II for more information on the action experiment More about Food). Although information tools such as 'Food Balance Sheets' might help, in addition, we need pricing strategies better able to reflect/incorporate environmental cost/benefits (e.g. subsidizing/taxing the environmental impact of food production processes). Here, there is a role for public government to address current market failures to include environmental costs and to open up consumer markets for a more inclusive, ecological economy. In addition, we need new business models which place 3P value creation and value distribution throughout the chain on a different footing.

Action experiments with new value distribution models show ways to open up existing market structures through introducing new production-distribution regimes. Central is a strategy aimed at a de-commodification of food through re-establishing the relationship between foodstuff on the one hand and the concrete food production process on the other (see Appendix II for more information on the action experiment Regional Food Chain – an action experiment actively organizing a new and more sustainable regional food market).

A second set of critical issues concerns the *institutional space for new forms of eco-efficiency*. By improving the substance sheet balance within current production processes, innovation practices run up to existing production chain divisions, both within and beyond current agribusiness, which are the result of a former production regime stressing forms of specialization. Partly, these divisions can be overcome. A clear example of this is represented by energy producing greenhouses (see also Chapter 2 and Appendix II for information on the action experiment SynErgie). Greenhouse horticulture can become more eco-efficient by greenhouses being better integrated in the energy supply system, thus enabling them to take their energy surplus to the energy market. In line with this kind of substance-chain innovation, other examples are bringing together different agro sectors within more integrated production environments, thus enabling a more efficient flow of substances between them. Here, one might think of horticulture making use of residues of pig breeding or dairy farming, and the other way around. In the course of time, this line of thinking, also addressed in terms of labels of industrial ecology, resulted in the development of business models aimed at the creation of new more integrated agri-business parks (so called *vital clusters*).

This line of thinking is rather clear and straight forward, also backed up by scientific models, both ecological and economic (see e.g. Smeets 2009). However, the actual development of such new, more integrated, production environments proves much more difficult than expected. Partly, problems can be related back to existing regime divisions, such as in terms of production cultures, the status of different production sectors amongst producers, the division of competences, the role of landownership, uncertainties about divisions of costs and benefits, et cetera. Up to a certain point, these kind of problems can be overcome by creating adaptive innovation environments, enabling a search for new economic models, also facilitating relations of trust and collaboration. However, in the end, investors back out if innovative initiatives become faced with serious problems in terms of their public legitimacy and/or access to land (e.g. forms of institutional anchorage). Basically, among others in the Netherlands, there is a serious struggle going on around notions and models of intensive/large scale farming. Here, the 'space of flow' within agriculture is meeting changes in the 'space of place' of rurality. Partly, this is a spatial problem, with the current farming landscape not being suited for a further enlargement of scale, due to the scale of the landscape, the lack of proper transport facilities, ground prices, etc. Partly however, this is also a cultural and political-normative issue, with more strict nature preservation policies, and landscape and animal welfare values/regulations challenging developmental opportunities. Here, we stumble upon increasing tensions accumulating on the interchange between 'the space of flow' and 'the space of places' in agriculture.

3.4. Where the space of flows meets the space of places

Studying transitions in agriculture from the perspective of the 'space of places', implies paying attention to the rapid changes the rural landscape and the related rural system is going through in the context of ongoing 'metropolitanization' processes.

First, there is the increasing significance of the regional scale in both spatial and economic policies. Because of the coming into being of the aforementioned 'network society', activities of companies and citizens are no longer concentrated in a common location or region. Technological and economic developments augment pressures on public space. As a result, at least in the Netherlands, the goals, effectiveness and legitimacy of spatial planning have lost much of their self-evidence. To resolve this issue, the importance of regional differentiation/ coordination and the involvement of regional stakeholders is stressed (WRR 1998). Besides, in economic terms, Porter (2003) led the way in pointing out how, due to the fact that there are substantial differences in economic performances across regions, the national scale less and less is a meaningful level of economic analysis. Determinants of economic performance should therefore be formulated at a regional level. As a consequence, in organizing innovations and transitions in agriculture in relation to its 'space of place', the regional level has replaced the national level as the critical entity.

At the same time, patterns of demographic growth, economic developments and the increase in physical and communicative mobility have resulted in radical changes in the formation of public space. Overall, in the Netherlands we see the coming into being of an interrelated space, sometimes described in terms of a North West European Metropolitan area, incorporating a stretch of land spreading from Amsterdam, over larger parts of the Netherlands, to Cologne and Brussels. In spatial terms, this area can be characterized as a 'network area', a landscape in which the city, nor the surrounding rural area, exists any longer as unambiguous phenomena (Boelens 2000). Virtual and physical networks connect residential, leisure and economic locations, events and people. Connectivity is the dominant factor instead of nearness.

As pointed out, this 'metropolitanization' processes also has a cultural dimension. Metropolitan norms and values start to dominate rurality. Metropolitan citizens differ from farmers with regard to their view on nature, landscape, agricultural production and animal welfare. Lörzing (2007) depicts this view as a demand for an 'idealized authenticity'. This involves landscapes with references to an idealized small scale and diversified rural history, such as Tuscany or the English countryside of Kent. Production activities and other references to techno-industrial activities are more and more evaluated as 'disturbances' (Braaksma and Bos 2007; Lörzing 2007).

In the Netherlands, this resulted in an increase in restrictive regulations for agricultural developments (Veldkamp et al. 2008). This change in the overall evaluation of rural landscapes increasingly started to conflict with the demands of the production regimes in which farmers had to operate. In the midst of the 20th century, urban, national and rural-agricultural regimes were aligned in a common outlook on the Dutch economic and spatial system. As such, agriculture functioned as an 'aligned interest', co-producing a favorable organization of space. By the end of the century, however, these different regimes more and more started to conflict with one another, producing a serious misfit between agriculture and the social organization of space. To illustrate this, we like to give a more detailed impression of developments in five significant action experiments.

3.4.1. Perspective from the space of places

A successful new regional development regime was created in the action experiment on care farming (see Appendix II for information on the action experiment Green Care). In this case different regimes were aligned: based upon their vicinity to a large urban demand, and their location in a favorable and protected national landscape, the farmers choose to increase their income by diversifying their farming activities. The professional approach which connected farmers, insurance companies and welfare organizations in a shared business model, played a decisive role. There was also a match between the economic interest of the farmers and local urban strategies: in recent years, the City of Amsterdam broadened its strategic outlook on the city from a strictly urban to a more regional, metropolitan approach, thus connecting urban and rural areas in a more integrated spatial perspective.

In contrast, a lack of success in realizing structural regional transitions expresses some interesting lessons as well. These can be traced back to various factors, in the end based upon the combination of a weak capacity to act in the face of stubborn institutional conditions. An action experiments situated in the hilly, small scale countryside in the south of the Netherlands shows an interesting example (see Appendix II for more information on the action experiment Heuvelland). The intention was to come to a new 'vital coalition' (a coalition able to bring about a regime-change), consisting of relevant stakeholders, jointly responsible for the conservation, maintenance and development of the landscape.

While actors indeed started to collaborate and individual projects did come about, in the end, the action experiment did not realize the regime change aimed for. Much energy was put in process aspects, to design and organize the networks which could act as the basis of a new coalition. However, the most important groups of actors in the project, the province, the entrepreneurs and the knowledge institutes, remained unable to align their interests productively. The province remained unable to overcome internal problems, such as administrative cleavages between its individual departments, solitary interventions of policy makers, and ongoing changes in developmental agenda's due to new elections. Knowledge institutes differed in their approach to the involvement of circles of stakeholder, due to different opinions with regard to the organization of regional transition processes. The entrepreneurs remained focussed on their own core-business, with a rather passive stand to regional development goals, which they treated as a prime responsibility for the province (Van der Stoep and Aarts 2010). In the meantime, the action experiment proved successful in the development of isolated business cases, such as the start up of new care arrangements between a medical hospital and rural hotels, and the creation of a central outlet for regional products in the provincial capital. Further on in the action experiment, the concept of Integrated Care Communities was developed (Haarman et al. 2009), mobilizing a health care organization, bank and project developers, and a housing company in the development of a new multifunctional context of community care.

In another action experiment, three agricultural businesses aimed at combining: an intensive pig breeding farm, a chicken farm and an installation company (see Appendix II for information on the action experiment New Mixed Farm). Together they would like to set up a joint bio-energy plant, thus relocating their business. Energy, waste flows and water will be exchanged. Thus, the participants developed new forms of cooperation as well as different types of sustainable, technological innovation (Horlings 2010; Smeets 2009). However, they ran up against profound misfits between their ideas and wider governmental and societal dynamics. In the end, 'society' has to provide the 'license to operate', the social legitimization for the new business model. However, social actors expressed a range of objections to the ideas, ranging from 'not-in-my-backyard' arguments concerning traffic pressures and the impact on the amount of particulate matter, to normative objections to the further industrialisation of intensive husbandry. The different governmental layers showed themselves sensitive to the public debate, and changed their originally supportive attitude and commitment to a more strict obedience to required spatial planning procedures, the public nuisance act and other regulations. The public discussion also led to the withdrawal of several potential horticulture partners. Entrepreneurs invested a lot in scientific evidence, but were not able to integrate this rational evidence into the normative public debate. In one way or another, the sustainable business storyline of the entrepreneurs did not align effectively with the normative and emotional storylines which dominated public debate.

In the action experiment on saline agriculture (see Appendix II for more information), the entrepreneur managed to create a successful business case by developing a market chain for saline crops from its production on the field to a number of high-ranking restaurants. The further development of saline agricultural was however blocked by the existing regimes of nature conservationists and agricultural organizations. Nature conservationists saw the development as a potential threat for existing or future saline nature areas. Agricultural organizations saw it as a threat because it could have a negative impact on the budgets spent on infrastructure to maintain sweet water agro-systems.

In another action experiment, the Northern Frisian Woods (see Appendix II for more information), a group of farmers started an experiment aimed at a radical regime change in national and European manure policy, the nutrient management experiment. The national Ministry of Agriculture and the scientists who were involved in drawing up the manure legislation opposed against the activities of the farmers. The farmers were, however, supported by another group of scientists, who argued for room for experimentation within the manure legislation. Together they wrote a letter to the minister asking for a manure experiment. After questions in national parliament, the minister agreed to an experiment. His successor granted a new room for experiment in 2010, despite negative advices of her advisors.

3.4.2. The importance of a 3P regional business case

What these five cases make clear is first of all that a successful regional innovation process should start with and keep a strong focus on the possibility of a business case. Without this focus, the energy of regional innovation processes leaks away to the process itself, and to organizational issues (like e.g. the public-private division of responsibilities). Hence, regime changes and the creation of vital coalitions are not to be regarded as aims in themselves but as an outcome of the exploration of new business cases. These business cases have to be developed by and based upon a self-organization of entrepreneurs, in collaboration with other organizations operating in informal and formal policy networks (e.g. Van der Stoep and Aarts 2010). An innovation initiative which starts with the design of networks and coalitions will lose much energy on organizational issues as misfits between stakeholders will not be bridged and networks will turn out to be fluid and instable. Alignments with wider groups of actors may be obstructed by various conditional circumstances, with a normative, emotional or rational background. Here, a powerful story line, supporting the business case, also in relation to its institutional environment, may help to create a fit.

In the case of the Northern Frisian Woods, a novelty received support from the minister, because a storyline[3] developed by farmers was made powerful by the support of scientists and members of parliament, despite it opposing existing legislation, and a negative advice by the existing regime (Van der Ploeg et al. 2006). Successful regional regime changes, such as has been the case in former Land Consolidation processes and in more recent nature preservation projects, had a strong business case orientation and a powerful storyline. An emergent business case with a powerful storyline and strong alignments with the institutional and social environment (thus combining economic, institutional and social forms of anchorage) may result in a vital coalition, aimed at realizing the value capturing potentials involved, not the other way around (Van der Stoep and Aarts 2010).

The economic, social and institutional anchorage or alignment of a storyline and the subsequent emergence of a vital coalition may be stimulated by versatile forms of leadership, a shared sense of urgency, or subtle and resilient support structures. But finally, as noted before, processes follow a more incremental line of events, with alternative storylines, and the business cases related, crystallizing out in different directions, in the end resulting in a victory of the most powerful and appealing ones (Geels et al. 2004; Horlings 2010). If subsequently these coalitions want to be the start of new regional development regimes, their storyline will have to be appealing to and accepted by wider societal groups. In order to reach this goal, people, planet and profit values and interests have to be included and in a certain sense feeding, targeting, and disciplining the story-telling process ('framing').

3.5. Conclusion: Breaking the developmental lock-in

Overlooking the landscape of innovation from the point of view of the interrelation between innovation dynamics and the wider institutional context, it first of all becomes clear that co-produced innovation programs in which collectively conducted action experiments take place, function as a platform for exploring and testing probable developmental trajectories. Taking a focused trial and error approach, organized around a strong learning based search for new triple P business models, with a clear focus on the interaction between practical and institutional characteristics, between process and content, they sensitize for innovative possibilities, and for ways of facilitating these.

At the same time, however, if operating effectively, they also make clear where market failures hinder normal innovation trajectories, thus focusing attention on the institutional conditions blocking a further transition process. Partly these institutional conditions may be drawn into the innovation program itself. For instance by formulating new action experiments with relevant stakeholders, thus taking institutional

[3] Chapter 4 further elaborates on the role of images in this process.

blockages up on the level of new innovation practices. This is especially the case where institutional blockages are situated in societal domains able to become part of this type of 'open' innovation programs. Think here of retailers opening up space for greener storylines, in collaboration with primary producers thus experimenting with alternative value capturing models. However, especially where institutional blockages involve government regulations and/or policies, which by their nature cannot be drawn into public-private programs as such, a different strategy is demanded for.

In line with this, looking through the complex field of relevant issues, social, techno-economic and institutional, there is one overarching topic which asks for attention. In abstract terms this involves the increasing misfit between techno-economic dynamics in the 'space of flow' of agriculture and socio-spatial dynamics in its 'space of 'places'. Within a common more abstract sustainable development perspective, places related dynamics on the one hand, stressing qualities of rurality, issues of food quality, animal welfare and biodiversity, and 'flow' related agricultural dynamics on the other, stressing issues of eco-efficiency, less and less align with one another, and are on a collision course.

This becomes particularly clear where innovation attempts try to go beyond existing production/planning regimes. As long as innovations adapt to the existing organization of supply chains, e.g. by focusing on an increase in eco-efficiency within existing production-distribution procedures, together with the spatial parameters related, this remains well within the realm of standard techno-economic reasoning. Here, there still are a lot of sustainability opportunities to be explored and exploited, with levels of success very much depending on technological opportunities and the cost-benefit dynamics related. However, innovations which attempt to go beyond the existing techno-economic system, focusing on a sustainable re-combination of supply chains, either within or beyond the agricultural system, easily run into problems. Partly these have to do with vested interests in the agro production system itself, with parties fearing for their position in a new and often yet uncertain field of competition, or with the fact that these new value chains are lacking alternative investment models, together with the appropriate chain management related. However, even where parties involved are showing a willingness to invest in alternative triple P business cases, both financially and organizationally, going beyond established value chains, they run into problems. In general terms these have to do with misfits between the alternative business propositions developed and the administrative 'landscape' within which these have to operate. A lot of issues have crossed the action experiments, from health, safety and insurance regulations, manure policy regulations and nature preservation regulations, to spatial/sectorial divisions of administrative responsibilities.

In more detail, two sets of problems stand out. Both have to do with the fact that new agricultural value propositions don't fit very well with the current administrative organization of (Dutch) space. On the one hand strategies aimed at a new eco-friendly generation of mixed, intensive agro business parks run up against a lack of both mental and spatial experimental space. On the other hand, more extensive, multifunctional farming strategies run up against a lack of comprehensive regional planning. Overarching is the problem of the non-existence of a strategic vision on the future of agriculture, formulating a stimulating developmental perspective, in line with the current metropolitan landscape, both in economic, in cultural and in spatial terms. For agriculture in high-densely populated areas to be able to take a next sustainability step, bridging the tensions between the ongoing internationalization of agricultural production and the increasing stress on a more sustainable quality of place, we have to search for a new fit between agriculture and space. In substantive terms, this perspective has been captured in the concept of Metropolitan Agriculture (see www.metropolitanagriculture.com). The concept covers a pragmatic attempt to come to a new, more sustainable, agricultural development logic or story line, focused on co-creating new connections between metropolitan areas and a wider field of agricultural developments. Thus the concept should create the physical, conceptual and institutional space to search for a new, more resilient relationship between the space of flows and the space of places in agriculture in the context of a densely populated but highly fertile area. Under such a developmental umbrella, there is room for a more heterogeneous agro economy, exploring a diversity of value proposition, both small scale and large scale, thus meeting the needs of a diversity of highly educated, highly mobile citizen-consumers, spanning the space between city and countryside.

In conclusion: if, within clearly defined sustainability targets, the administrative environment would allow itself the space to experiment with new conceptions of the interrelationship between agriculture and the metropolitan environment, it would at the same time unleash the kind of institutional space enabling a productive search for more sustainable forms of system innovation. Together, this forms a clear illustration of how agro innovations need the kind of responsive institutional environment to enable exploring real system innovation. A clear case of the necessity for an ongoing circular interaction between experimentation and context, niche and landscape, process and content.

3.6. References

Andersson, K., Eklund, E., Lehtola, M., & Salmi, P. (2009). Beyond the rural-urban divide. Cross-continental perspectives on the differentiated countryside and its regulation. *Research in rural sociology and development*, vol. 14. Bingley, UK: Emerald.

Beck, W. (1986). *Risikogesellschaft. Auf dem Weg in eine andere Moderne*. Frankfurt am Main: Suhrkamp Verlag.

Boelens, L. (Ed.) (2000). *Nederland Netwerkenland, een inventarisatie van de nieuwe condities van planologie en stedebouw*. Rotterdam: Nai Uitgevers.

Braaksma, P.J., & Bos, A.E. (Eds.) (2007). *Investeren in het Nederlandse landschap. Opbrengst: geluk en euro's*. Den Haag: Ministerie van LNV.

Bunte, F., Bolhuis, J., Bont, J.C. de, Jukema, G., & Kuiper, E. (2009). *Prijsvorming van voedingproducten*. Nota 09-074. Wageningen: LEI.

Castells, M. (1996). *The rise of the network society*. The information age: economy, society and culture, vol. I. Cambridge, MA: Blackwell.

Castells, M. (2009). *Communication power*. Oxford: Oxford University Press.

Geels, F., Schot, J., & Verbong, G. (2004). *A multi-level analysis of historical transitions in agriculture and food.* Position paper for Bsik programme. Zoetermeer: TransForum Agro & Groen.

Geels, F., Elzen, B., Berkers, E., Leeuwis, C., & Mierlo, B. van (2008). *Historische en hedendaagse systeem-innovaties in de glastuinbouw en varkenshouderij. Een innovatie-sociologische analyse.* Zoetermeer: TransForum Agro & Groen.

Geels, F.W., & Schot, J. (2007). Typology of sociotechnical transition pathways. *Research Policy,* 36, 399–417.

Grin, J., Rotmans, J., & Schot, J. (2010). *Transitions to sustainable development.* New York: Routledge.

Haarmann, W., Horlings, I., & Derix, G. (2009). *Please in my backyard. Parc Hoogveld en de opkomst van de Integrated Care Community. TransForum Working Paper 10*. Zoetermeer: TransForum Agro & Groen.

Hassink, J., Zwartbol, Ch., Agricola, H., Elings, M., & Thissen, J.T.N.M. (2007). Current status and potential of care farms in the Netherlands. *Netherlands Journal for Agrarian Sociology, 55* (1).

Horlings, I. (Ed.) (2010). *Vital coalitions, vital regions. Partnerships for sustainable regional development*. Wageningen: Wageningen Academic Publishers.

Hoogma, R., Kemp, R., Schot, J., & Truffer, B. (2002). *Experimenting for sustainable transport: The approach of strategic niche management.* London and New York: Spon Press.

Klerkx, L., Aarts, N., & Leeuwis, C. (2010). Adaptive management in agricultural innovation systems: the interactions between innovation networks and their environment. *Agricultural Systems*, *103*(6), 390–400.

Krasner, S.D. (Ed.) (1983). *International regimes*. Ithaca, NY: Cornell University Press.

Leeuwis, C., & Aarts, M.N.C. (2008). *Conflict, communication and space for change in complex problem situations.* Paper presented at the 2008 conference of the International Association of Conflict Management (IACM), Chicago, July 3–6, 2008.

Loorbach, D. (2007). *Transition management: new mode of governance for sustainable development.* Dissertation. Rotterdam: Erasmus University.

Lörzing, H. (2007). Landschap, ons themapark. In: Derksen, W. (Ed.), *Kijkend naar Nederland* (pp. 43–49). Rotterdam/Den Haag: Nai Uitgevers/Ruimtelijk Planbureau.

Marsden, T. (1998). New rural territories: Regulating the differentiated rural spaces. *Journal of Rural Studies*, *14*, 107–117.

Mommaas, H. (2003). Op zoek naar Arcadie. Over de sociaal-culturele dimensie van de groene ruimte. In: Beun, N.J., Hulshof, C.W.M., & Hillebrand, J.H.A. (Eds.), *Sociaal-culturele aspecten van groene ruimte en voeding* (pp. 47–81). Rapport nr. 03.2.034. Den Haag: Innovatienetwerk Groene Ruimte en Agrocluster.

Mossberger, K., & Stoker, G. (2001). The evolution of urban regime theory. The challenge of conceptualization. *Urban Affairs*, *36*, 810–835.

Nederlandse Mededingingsautoriteit (2009). *Prijsvorming in de agri-food sector*. Den Haag: NMA.

Nooteboom, B., & Stam, E. (2008). *Micro-foundations for innovation policy*. Amsterdam, Amsterdam University Press.

Van der Ploeg, J.D., Verschuren, P., Verhoeven, F., & Peppels, J. (2006). Dealing with novelties, a grassland experiment reconsidered. *Journal of Environmental Policy and Planning*, *8*(3), 199–218.

Poppe, K.J., Termeer, C. & Slingerland, M. (Eds.) (2009). *Transitions towards sustainable agriculture and food chains in peri-urban areas.* Wageningen: Wageningen Academic Publishers.

Porter, M.E. (2003). The economic performance of regions. *Regional Studies, 37*(6&7), 549–578.

Prigogine, I., & Stengers, I. (1984). *Order out of chaos: Man's new dialogue with nature*. New York: Bantam Books.

Ruggie, J.G. (1975). International responses to technology – concepts and trends. *International Organization*, *29*(3), 557–583.

Schot, J., & Geels, F.W. (2008). Strategic niche management and sustainable innovation journeys: theory, findings, research agenda and policy. *Technology Analysis & Strategic Management, 20*, 537–554.

Smeets, P. (2009). *Expeditie Agroparken.* Dissertation. Wageningen: Wageningen UR.

Van der Stoep, H. & Aarts, M.N.C. (2010). New markets in Heuvelland: coalition building and agenda setting. In: Horlings, I. (Ed.), *Vital coalitions, vital regions. Partnerships for sustainable, regional development* (pp. 121–150). Wageningen: Wageningen Academic Publishers.

Stone, C. (1989). *Regime politics.* Lawrence, KA: University Press of Kansas.

Stone, C. (1993). Urban regimes and the capacity to govern: a political economy approach. *Journal of Urban Affairs, 15*, 1–28.

Stoker, G., & Mossberger, K. (1994). Urban regime theory in comparative perspective. *Environment and Planning C: Government and Policy, 12*, 195–212.

UN (2006). *World Population Prospects. The 2004 Revision.* Volume III: Analytical Report. New York: United Nations Publication.

USDA Foreign Agricultural Service (2005). *Netherlands agricultural situation, the Benelux horticulture market.* GAIN Report Number NL5002. Retrieved 2010-06-07.

Wetenschappelijke Raad voor het Regeringsbeleid (WRR) (1998). *Ruimtelijke Ontwikkelingspolitiek.* Rapporten aan de Regering. Den Haag: SDU Uitgevers.

Veldkamp, A., Van Altvorst, A.C., Eweg, R., Jacobsen, E., Van Kleef, A., Van Latesteijn, H., Mager, S., Mommaas, H., Smeets, P.J.A.M., Spaans, L., & Van Trijp, H. (2008). Triggering transitions towards sustainable development of the Dutch agricultural sector: TransForum's approach. *Agronomy for Sustainable Development, 28*, 87–96.

Chapter 4. For or against innovation? The influence of images

Pieter J. Beers[1] **and Tom Veldkamp**[2]

[1]PJ.Beers@wur.nl; TransForum and Education and Competence Studies, Wageningen University, The Netherlands

[2]Veldkamp@itc.nl; Faculty ITC, University of Twente, P.O. Box 6, 75AA Enschede, The Netherlands

Abstract Sustainable development can be regarded as a wicked problem. Consequently, standard problem solving strategies will not work, but processes requiring engagement and connection of different stakeholders may yield solutions toward a more sustainable future. Norton (2005) proposes an adaptive ecosystem management approach to deal 'better' with sustainability by adhering to three principles: Experimentation, Multi-scalar analysis, Place sensitivity. In order to allow experimentation, and to overcome closed-form definitions of sustainability goals, images are crucial to mobilize stakeholders in innovation. An image mobilizes stakeholders towards a shared innovation effort. This chapter explores the role of images within innovations and societal images that influence innovation. We suggest options for dealing with images: collaborative visualization and image management.

4.1. Introduction

'Who is the farmer?', such an innocuous question, but such a wide range of answers. That is probably why, in practice, we will give simple answers most of the time: The farmer is 'the custodian of the rural landscape', or 'the feeder of the world'. Not a complete answer, but rather an image of what we really mean: metaphorical, simple, and easily communicated. And this simplification is not made without reason, because the underlying complexities are often too comprehensive to communicate. For instance, stating that the farmer is the custodian of the countryside also means that his/her cows must be in the pastures. It implies small-scale farming businesses, and ascribes important value to the landscape as a socio-cultural resource.

Images have strong, mobilizing qualities, because they are often associated with very basic value orientations. For instance, public resistance against genetically modified organisms (GMOs), irrespective of whether it is justified, appears much stronger than warranted on the basis of scientific insights on its risks for biodiversity

H.C. van Latesteijn and K. Andeweg (eds.), *The TransForum Model: Transforming Agro Innovation Toward Sustainable Development*, DOI 10.1007/978-90-481-9781-1_4,

alone: food from GMOs is seen as 'Frankenfood'. Other typical examples of the strong mobilizing power of images are biofuels, which were first seen positively, as a solution for climate change, and later on negatively, as a cause of world food shortage.

We often encounter images where opposing value orientations meet in public debate. Opposite values like animal welfare versus animal production (Hurley 2008) or organic agriculture versus intensive agriculture (cf. Borch 2007; Hurley 2008) result in the use of simple images to frame associated innovations as either good or bad. For instance, animal welfare values lead to branding industrial farming with the image of a pig factory and connotations of disease-outbreaks (e.g., avian flu, mad cow disease) with images of cattle eradication. And the converse: animal production proponents use the image of tree huggers to frame organic farming as cute, but not very effective. One and the same innovation can be a clear-cut image of progress to entrepreneurs, and a nightmare for environmentalists.

Both the power and the abundance of images lend credence to the belief that images can make or break an innovation. Images continuously influence the innovative potential of promising inventions. And moreover, images are constantly changing: there is no 'true' image. Their mobilizing potential may be at the heart of non-linear developments in system innovation, and the multi-stakeholder requirements for such innovation imply that images not only act as a societal influence 'out there', but also as a given within innovation. When stimulating or working on innovation, it is important to know both how images may affect innovation, and how images can be managed to improve the innovative potential of an initiative. However, hardly any scientific analysis addresses this phenomenon.

This chapter therefore focuses on two questions:

1. What is the role of images in agricultural innovation for sustainable development?
2. How can images be managed to increase the success of an innovation?

We first discuss some examples of the role of images from scientific literature. We then introduce our research framework. The chapter proceeds with a study of the role of images in two selected action experiments. We conclude with recommendations for collaborative visualization and managing images.

4.2. Images in innovation

Working on innovation toward a more sustainable future requires us to explore the future. This often includes constructing new value propositions of the innovation and, if successful, ends with putting the innovation into practice. This process often includes an image of the now (how things are) and a vision of the future (how things could and should be), which can be seen as an idealized image, with normative connotations (Van der Helm 2009).

For reasons of clarity, it is important to disambiguate the concepts of *image* and *vision*. In this chapter, we define an image as a simple, metaphorical representation of a complex real-world phenomenon (Beers et al. 2010). A metaphor is a powerful heuristic because it conveys complex reality in a simple, recognizable and credible way. The simplicity of images can be seen as a response to the complexity of what is portrayed – on the one hand, a simple image enables communication about complex subject matter that otherwise would not be discussed, but on the other hand the same simplicity makes it hard to track how it reflects a complex reality. Finally, images are often value-laden but they are convincing for the stakeholders involved.

Visions are defined as:

> The more or less explicit claim or expression of a future that is idealized in order to mobilize present potential to move into the direction of this future. (Van der Helm 2009: 100)

Defined this way, both 'vision' and 'image' have the potential to mobilize (groups of) people, both are aimed at (persuasive) communication, and both have the character of a social representation. In other words, both images and visions are metaphors. The main difference is that visions necessarily concern the future, which need not be the case for images. Another difference is that images are not necessarily idealized. A vision is meant to depict a desirable future, which can act as guiding principle for decision making. An image does not necessarily depict something as desirable. An image may just as well depict something as undesirable, to mobilize others against certain actions. In sum, visions can be seen as a subset of images, in the sense that a vision is an image of a desirable future.

In multi-stakeholder processes,[1] images can become the nexus for normative tensions between different value orientations. For example, Hurley (2008) sees the countryside as vulnerable, in need of protection from urbanization, whereas others regard urban functions for rural areas as the salvation of the countryside. And Alexandra and Riddington (2007) juxtapose the image of nature as 'the provider of life-fulfilling ecological services' with the image of nature as 'a cornucopia of resources available for exploitation'.

Similar images have a long history in marking differences between discourses about the rural area (Frouws 1998). Some see the farmer as the custodian of the rural landscape, which fits well with small-scale agriculture, family farms, and a sizeable role for organic agriculture. Others see the family farm as a relic of a bygone age, ill at ease with the entrepreneurial role of the farmer as the world provider of food. And again others adhere to the image of a multifunctional rural area, in which agricultural functions are mixed with recreation, biodiversity and health care. Simple, iconic images serve as a focal point for public debate between different discourses about rural areas. No single set of images, no single discourse has established dominance over the other discourses. Instead, different discourses

[1] Multi stakeholder processes are required for system innovation, see Chapter 1.

co-exist, in ongoing interaction and conflict, conflict that is symbolized and empowered by images.

Innovation itself is societally embedded. As a consequence, many inventions and innovations have become associated with specific images, and met with strong societal pressure (Hoes et al. 2009). An innovation can only succeed when it has sufficient societal support. In other cases, simple images have had a decisive influence on the content and goals within an innovation. However, analyzing complex systems, such as sustainable development and system innovation, is especially hard because of their non-linear and counter-intuitive behaviors (Johnson 2008; Kay et al. 1999). While mobilizing strong support and swift implementation, images may fail to do justice to the complexity of the issue at hand, not leading to the desired outcomes.

4.3. Hypotheses and research approach

4.3.1. Hypotheses

From the above, we conclude that images play a role both within an innovation project, and on the societal interface of an innovation. With an innovation project, we mean in this chapter a process in which multi-stakeholders aim to establish an innovation in practice (such as the action experiments conducted in the context of this book). Images, however, have multi-scale dynamics which need to be addressed. A successful mobilizing image within an innovation project might trigger strong opposing pressure at the societal level.

Within an innovation project, images are continuously shaped and modified, interacting with processes of problem analysis and the construction of innovative technologies and practices. Images can decisively influence the content and direction of an innovation project, by affecting its innovative potential. They can act as the catalyst for starting or closing a project, by offering a simple, accessible view of the underlying complexity. However, if the image in question turns out to be inaccurate, the project's goals and main assumptions may be unrealistic as well and the image needs to be modified.

In society, images are embedded in discourses on the various roles of agriculture, and concern the farmer, various types of farming (extensive vs. intensive, organic vs. industrial), and innovations. The role of images comes to the fore on the interface between an innovation project and society. Different (societal) groups use and produce different images of agriculture and of innovation. When an innovation project encounters an adversarial societal group, such a meeting can results in a clash of images, where the project can only rarely change the societal image. We derive two main hypotheses from the above:

1. Images of sustainability enhance the innovative potential of an innovation project when they convey complex reality in a manner that is simple, credible, understandable and yet reflective of that reality.
2. In societal debate, images influence the innovative potential of innovation projects.

4.3.2. Research approach

Our research approach is summarized in Figure 4.1; images exist both within an innovation project and at the societal level. In society, images are embedded in discourse (represented by the clouds in Figure 4.1). The role and influence of images is depicted by the arrows in Figure 4.1. The role of images can be studied on two levels: within an innovation project and on the interface between the project and society.

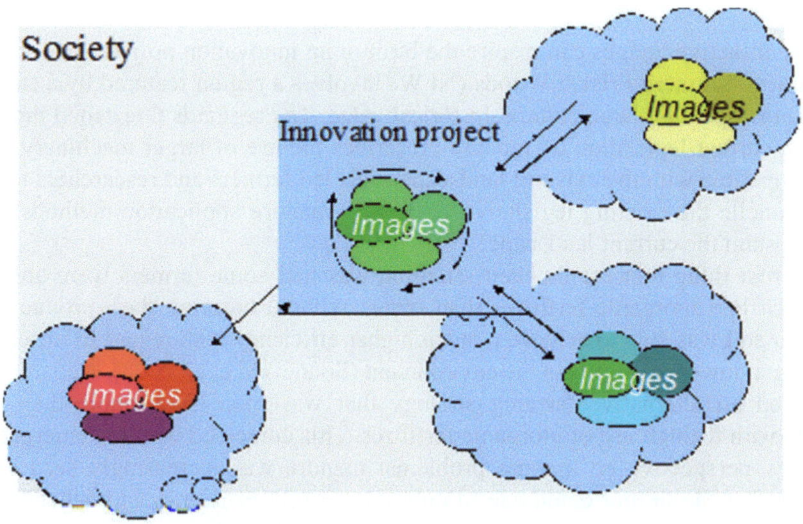

Fig. 4.1. Framework for studying images.

Within innovation projects, images are hypothesized to be a major factor in the project's potential for innovation. Images offer an important reduction of complexity. Because sustainable development is a wicked problem no one knows the real complexity, requiring images to be adaptive and stakeholders to be willing to work on a learning-by-doing basis. Our hypothesis regarding the societal level pertains to the interface between (civil and political) society and innovation projects, and how images affect an innovation project. This hypothesis requires studying to what extent an invention has the potential to be adopted as an innovation, and whether this adoption is influenced by images at the societal level.

It requires a focus on societal discourse (which itself embeds specific images) and actors. We studied the role of images on the level of innovation projects and on the interface between the project and society.

4.4. The role of images – case studies

We studied the role of images in two action experiments. We explored the Northern Frisian Woods action experiment to highlight the possible roles of images within an innovation project, and the New Mixed Farm action experiment to explore the influence of societal images on an innovation. Appendix II of this book gives more information on these action experiments.

4.4.1. Images within innovation projects – Northern Frisian Woods

Simple, attractive images can inspire the birth of an innovation project. The action experiment Northern Frisian Woods (NFW) involves a region featured by a small-scale landscape with large cultural-historical value. The region is threatened though because current legislation on manure prescribes the use of larger machinery that is incompatible with the existing landscape. This led farmers and researchers to try and reconcile the existing legislative goals with manure application methods that could sustain the current landscape.

The first thing that caught their attention was that some farmers were able to use much less inorganic fertilizer than others without harming their production, and their soil was able to achieve a much higher efficiency with regard to nitrogen, resulting in lower emissions to the environment (Sonneveld et al. 2008). Apparently, they used an alternative farming strategy that was more sustainable than the default, with its high use of inorganic fertilizer. This coincided with the emergence of a new perspective on manure problems: manure was increasingly seen as a resource that the farmer could use to improve nutrient efficiency in his/her farm, instead of an environmental problem (Stuiver 2008). These observations and developments gave rise to the idea that the farmer's management strategy could lead to more sustainable fertilizer use. In terms of a simple image: *the land (and environment) adapts to the farmer*.

At the same time, local actors came to see the legislative situation in terms of David and Goliath (Stuiver 2008): the farmers (David) were to quest for changes in the existing manure legislation (Goliath). Indeed, existing legislation was the only reason that prevented more widespread experimentation with the alternative management style: it used a fertilizer application method that was not legally allowed, despite its lower nitrogen emissions to the environment. The project could only start with an exemption from manure legislation. In this situation, the simple image that the land adapts to the farmer, combined with the advent of a movement towards change, together appear to have generated sufficient pressure to sway the legislators. The exemption was granted, and a 6-year project to study the effects of the alternative strategy got under way.

In a scientific systemic study, the accuracy of this image was tested (Van Apeldoorn et al. Submitted) by applying a mathematical computer model based on complex adaptive systems theory (Holland 1996; Kay 2002). The complexity perspective used in the model included concepts like resilience, adaptability and transformability (Holling 1973; Scheffer et al. 2001). From the existing diversity in landscape and soil types, it became clear that soils rich in organic matter, with high nitrogen efficiency and low environmental impacts, coincided with the alternative management strategy. Modeling experiments demonstrated that a transition from the default to the alternative style indeed is possible, but not within the time-frame perceived by the project members (about 7 years). Instead, modeling results indicated that it would take nearly 200 years to achieve the set goal. Contrary to the initial image, the alternative strategy did not control soil nutrient efficiency. Instead, farmers appear to have adapted their management to the soil composition of their pastures, in other words, to their location in the landscape. This leads to a new image 'the farmer adapts to the land'. Currently this new image has not been used by the action experiment yet, but they have discovered that the results of their image are not as expected (Stuiver 2006).

From this account we can derive two important lessons about images. The first is that strong images can play an important role in raising awareness of an innovation potential. In the case of the North Frisian Woods, a simple image helped mobilize a sustained effort to convince government officials to allow an exemption that enabled starting an innovation project. This illustrates the mobilizing potential of a simple image.

The second lesson is that a simple image can foster the potential of an innovation project, but only if it is reflective of the reality that it portrays. In the NFW example, based on short term evidence, images gave rise to the impression that soil efficiency could be easily adapted. A closer look at the complex system dynamics involved revealed that this is not the case after all. The image that initially mobilized the support to petition for an exemption from the law did not realistically reflect how long it would take to switch to the alternative strategy. As a result, the innovative potential of the Northern Frisian Woods action experiment is at stake, because the exemption that is necessary to prolong the experiment is based on the expectation that it is possible to switch to the alternative strategy within a timeframe of about 7 years.

We conclude that simple images were initially instrumental to harness the innovative potential of the action experiment, by mobilizing a movement to get an exemption. However, the same image did not yield the expected results because the role of the underlying system's complexity was underestimated. The next step in solving this wicked problem could be image adaptation including these new insights.

4.4.2. Societal influences on innovation experiments – New Mixed Farm

The New Mixed Farm (NMF) action experiment provides an account of how societal forces can influence an innovation and the role images can play in such a case.

The environmental advantages to the NMF are manifold. Firstly, waste flows of nitrogen and phosphorus are reduced because of manure reuse and biogas production from manure. Secondly, the co-location of animal breeding and a slaughterhouse reduces the amount of traffic around the farm, which in turn reduces the risk of animal disease transmission. Thirdly, by locally producing energy from waste, large reductions in CO_2 emission are established. In sum, the NMF can be seen as a more environment-friendly version of traditional industrial farming.

Notwithstanding its environmental advantages, the NMF met with large adversity in the public debate, which posed a threat to its innovative potential. Why? The NMF raised many public concerns with regard to animal welfare, increases in local animal transports, environmental pollution, and spread of animal disease. In the eye of the public, the NMF had become the arch-image of industrial farming, taken to the next level. Subsequently, all concerns that usually hold for industrial farming got attributed to the NMF. The NMF came to be seen as almost the opposite of what it actually was, especially regarding its environmental impacts.

The NMF-example can be seen as a clash between two discourses. On the one hand, the action experiment itself has its origin in a sustainability-oriented utilitarian discourse (Hermans et al. 2010), which emphasizes the role of technological solutions to sustainability issues. The regional context of the action experiment clearly was adverse to the discourse of traditional (not sustainability oriented) utilitarian agriculture. As the NMF resembles agricultural developments that belong to this discourse, it quickly became seen as yet another example of unsustainable utilitarianism. On the other hand can we also interpret the conflict as a multi-scale image clash. The local stakeholders initially all agreed on the image of the NMF, but this image was attacked at the national level by stakeholders with different national stakes.

From this account, we can draw a number of lessons about images. Firstly, it provides another example of the mobilizing power of simple images, in this case *against* the innovation in question. In this case, a negative societal image came to threaten the innovative potential of the NMF. When the image was developed in a multi-scale context, mobilizing support at local and national level, this would probably not have happened. Secondly, images obscured many of the characteristics of the actual innovation. The characteristics of industrial farming became superimposed on the NMF, while many of those characteristics were quite the opposite of the actual characteristics of the innovation. In other words, the metaphorical content of the image of industrial farming became more prevalent than the actual content of the innovation, which in turn means that the societal image of the innovation was not reflective of the innovation itself.

We conclude that simple images can mobilize societal support against an innovation, and that, in some cases, the metaphorical content of an image can stand in the way of its reflection of reality. In the worst case, an innovation's image may prevent it from getting implemented at all. This is a call for ways to manage images, starting with the awareness of one's image, and continuing with the projection of the image an invention or innovation aspires to. Reflexive awareness of its own images, and how they are received societal, can facilitate an innovation in adapting to existing societal discourses and the images embedded in them, which in turn can foster its innovative potential.

4.5. Dealing with images in innovation

Knowing the importance of images for innovation, how can we deal with their associated potential and problems? Can we derive from our analysis a perspective on how to deal with images? Below, we draft some tentative directions for dealing with images in innovation. We propose that collaborative visualization and image management can be used as a way to deal with images in innovation toward sustainable development.

4.5.1. Collaborative visualization

In innovation trajectories, the accuracy of simple images needs to be thoroughly tested, and the assumptions inherent to those images need to be regarded with healthy skepticism. Visualization can help doing this. By visualization, we mean the act of making something visible, making an image. Using the right visualization technique, implicit assumptions can be made explicit and tangible to others. Resulting images ease the communicative processes, especially when the system complexity exceeds the communicative bandwidth shared by the stakeholders in the team – as the saying goes, a picture is worth a thousand words. The trick is to choose a visualization technique that is appropriate for the problem involved – for instance, when buying a car, it does not pay to visualize its inner workings if you're interested in its color.

In the Northern Frisian Woods example, we strengthened our system analysis by using theory about complex adaptive systems. This suggests that, for cases like the NFW, it helps to use visualizations based on complex adaptive systems. In such a way, carefully chosen visualization techniques can help to increase the accuracy of the resulting images.

Complexity is one of the hallmark characteristics of systems innovation and transition challenges. Taking the role of complexity into account means that it must be *visualized* just as simply as other images would be, while doing justice to the underlying content. It is not hard to imagine how some of the more problematic aspects of complexity might be visualized in a simple, accessible manner (see Figure 4.2). For instance, non-linear shifts in complex systems can be easily visualized by a sharp change in the slope of a curve. Robustness is often visualized with the 'ball-in-a-cup' metaphor; the deeper the cup, the lower the chances of the ball leaping out of the cup, and the more robust the system.

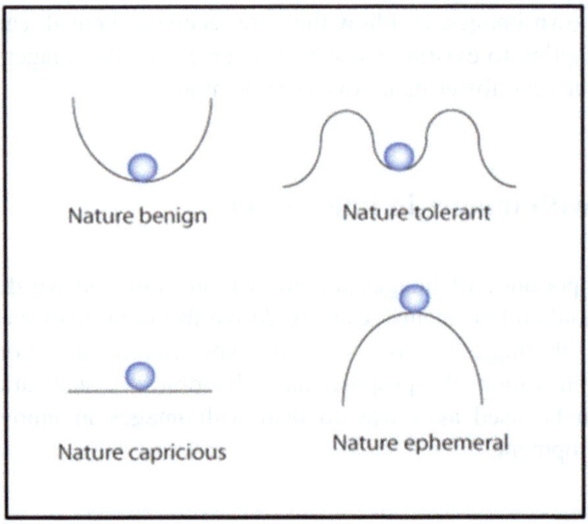

Fig. 4.2. Simple example visualizations of Myths of Nature (adapted from Schwarz and Thompson 1990).

Visualization can be used to improve the extent to which images are reflective of the underlying system's complexity. Specific visualizations are chosen for their reflective accuracy, not their mobilizing power. The ball-in-the-cup-metaphor thus can be used to visualize a system's robustness. The assumptions about systemic complexity that are inherent to an action experiment can be explored by using such visualizations. Collaboratively visualizing the future can thus be used to establish a more thorough understanding of the systems involved. This way, an experiment can escape false assumptions inherent to simple images.

While no established technique of collaborative visualization currently exists, we certainly can imagine ways in which it can be used in practice. An innovation project often starts with a simple, mobilizing image. Next, the participants need to explore the implicit assumptions that are inherent to that image. The participants in the project can use collaborative visualization to map out their implicit assumptions about the underlying systems. They use specific visualizations that are tailored to accurately reflect complexity.

For instance, the image that the land adapts to the farmer (see the NFW example) has the inherent assumption that there are no structural obstacles to changing from one farm management strategy to another. In this case, collaborative visualization means that the participants make scenarios about the future of the underlying system, and the effects of the action experiment. They use visualizations that help to make assumptions about the future and about time scales explicit. Instead of blindly going with the implicit assumptions inherent to the issue, the scenario helps the participants to analyze the innovative potential of the project by thinking about the future.

Currently, efforts are under way to use visualization in the context of collaboratively making a regional development scenario (Vervoort et al. 2010). Our current working hypothesis is that collaborative visualization techniques can enhance the faithful treatment of complexity in innovation, and that they enable creating accessible images of complex phenomena that are more reflective of the complex aspects of reality. Ultimately, we aim to use visualization to enhance the innovative potential of innovation projects. The proposed visualizations will also allow quick updates of adaptive images when new insights are gained in the projects.

4.5.2. Managing images

It is not only the experiment's content that counts on the interface between an innovation project and society at large. Images can be seen to act as sets of boundary objects (Regeer 2010). In some cases, the process in which a project becomes associated with societal images – be it positive or negative – appears to a large extent autonomous, outside the control of the experiment's participants. It would seem that innovation projects are subject to continuous framing and reframing activities (Schön and Rein 1994), which in some cases do little justice to a project's content. How can we adapt innovation's images to existing regime forces, in order to secure their innovation potential?

Our conclusions about the New Mixed Farm show the importance of reflexive awareness of images, and of the societal reception of the action experiment. Reflexive awareness may help an innovation project to adapt to existing societal images, which in turn can foster the project's innovative potential. One way in which this may be done is to engage societal actors early-on in the development of an innovation project in order to acquaint oneself with the various ways the project may come to be seen. This recommendation substantiates the sustainable development assumption that system innovation requires a multi-stakeholder approach (Veldkamp et al. 2008). Engaging multiple societal stakeholders can increase the extent to which multiple existing discourses are represented within an innovation project, which increases the project's awareness of potential clashes with societal images.

Our experiences with the NMF action experiment indicate more ways for a project to increase its awareness of societal images and their influence. The first suggestion is that existing relations between an innovation project and discourses must be monitored and modeled. That way, a possible conflict between the innovation's content and the way the innovation experiment is perceived, its image, can be detected earlier on, before it has resulted in substantial public protest. Hoes et al. (2009) suggest that, to curtail public protest, public actors need to be included in innovation projects as early as possible. It appears that multiple societal views need to be engaged to be able to create images. This confirms the fifth assumption for sustainable development, that multi-stakeholder approaches imply trans-disciplinary knowledge creation.

Discourse analysis, as a social sciences research method, can yield a more comprehensive view of existing discourses and their associated images (e.g., Frouws 1998; Hermans et al. 2010), but it is also more labor-intensive, which may not be feasible in the context of an innovation.

Having established the various relevant societal images, and which are positive and which are negative, the second step towards managing images is to actively frame the innovation favorably, using those societal images that are positive. For instance, one and the same invention or innovation can be described both by its technical merits, and by its environmental effects. In the case of agriculture, the former may lead to a negative image (the food isn't natural), whereas the latter puts the experiment in a positive light.

The case of cisgenesis (see Chapter 2) is a good example of using positive images to frame an invention. Technically speaking, cisgenesis does not differ from transgenesis, a technique of genetic modification. In the case of food production, transgenesis has got the image of 'Frankenfood'. It stands to reason that cisgenesis runs the risk of becoming associated with the same negative image. However, cisgenesis differs from transgenesis in that it only transplants genetic material *within* one species. This means that the resulting genetically modified organisms could also have resulted from cross-breeding. It only happens faster. Framing cisgenesis as 'cross-breeding, but faster' highlights the difference between transgenesis and cisgenesis, and also explains that the environmental risks associated with cisgenesis are much lower than those associated with transgenesis.

4.6. Conclusion

In the face of sustainability problems, experiments are needed to start innovation. In doing so, organizers of innovation get confronted with societal images, that influence an innovation's potential. Noting the importance of images, we raised the question of the role of images in agricultural innovation for sustainable development, and how these images can be managed to increase the success of an innovation.

Our analysis of the Northern Frisian Woods showed that images of sustainability enhance the innovative potential if and only if they convey complex reality in a manner that is simple, understandable, credible and yet reflective of that reality. In the NFW, an image initially supported the action experiment, but later on that same image required adaptation, because the image was not sufficiently reflective of the underlying system's complexity.

The cases also exemplified how societal discourses can affect the potential of an innovation by affecting their image. Our results show that it is very important how an innovation is viewed: An innovation's content may radically differ from the images people have, and when associated with negative images, the innovation's potential decreases. This result suggests that innovation projects need not only focus on content, they also need to be careful of their image.

The first assumption to enhance innovation toward sustainable development, as addressed in Chapter 1, claiming sustainable development is a dynamic process, underlines the risks associated with the simplicity of images; the dynamic process associated with (un)sustainable development are easily overlooked with simple images. The only way to deal with this dynamic process is the ability to adapt the image when learning new insights during the innovation project. The results also show that images are involved in the non-linearity of system innovation. The technical aspects of system innovation cannot be developed linearly, because the emergence of images can cause sudden changes in their societal reception. It follows that an innovation project itself needs to prepare for non-linearity, which confirms the principle that system innovation is a non-linear learning process.

4.7. References

Alexandra, J., & Riddington, C. (2007). Redreaming the rural landscape. *Futures, 39*, 324–339.

Beers, P.J., Veldkamp, A., Hermans, F., Van Apeldoorn, D., Vervoort, J.M., & Kok, K. (2010). Future sustainability and images. *Futures, 42*, 723–732.

Borch, K. (2007). Emerging technologies in favour of sustainable agriculture. *Futures, 39*, 1045–1066.

Frouws, J. (1998). The contested redefinition of the countryside. An analysis of rural discourses in The Netherlands. *Sociologia Ruralis, 38*(1), 45–68.

Hermans, F., Horlings, I., Beers, P. J., & Mommaas, H. (2010). The contested redefinition of a sustainable countryside; revisiting Frouws' rurality discourses. *Sociologia Ruralis, 50(1)*, 46–63.

Hoes, A.C., Regeer, B., & Bunders, J. (2009). *Anticipating public protest in niche experiments.* Paper presented at the First European Conference on Sustainability Transitions June 4–6 2009: Dynamics & Governance of Transitions to Sustainability, Amsterdam, The Netherlands.

Holland, J.H. (1996). *Hidden order: How adaptation builds complexity.* Reading, MA: Addison-Wesley.

Holling, C.S. (1973). Resilience and stability of ecological systems. *Annual Reviews of Ecological Systems, 4*, 1–23.

Hurley, K. (2008). Food in the future: Does futures studies have a role to play? *Futures, 40*, 698–701.

Johnson, J. (2008). Science and policy in designing complex futures. *Futures, 40*, 520–536.

Kay, J.J. (2002). On complexity theory, exergy, and industrial ecology: Some implications for construction ecology. In C. Kibert, J. Sendzimir, & B. Guy (Eds.), *Construction ecology: Nature as the Basis for Green Buildings* (pp. 72–107). London: Spoon Press.

Kay, J.J., Regier, H.A., Boyle, M., & Francis, G. (1999). An ecosystem approach for sustainability: Addressing the challenge of complexity. *Futures, 31*, 721–742.

Norton, B.G. (2005). *Sustainability: A philosophy of adaptive ecosystem management*. Chicago: University of Chicago Press.

Regeer, B.J. (2010). *Making the invisible visible. Analysing the development of strategies and changes in knowledge production to deal with persistent problems in sustainable development*. Oisterwijk: Boxpress.

Scheffer, M., Carpenter, S., Foley, J.A., Folke, C., & Walker, B. (2001). Catastrophic shifts in ecosystems. *Nature, 413*, 591–596.

Schön, D.A., & Rein, M. (1994). *Frame reflection. Toward the resolution of intractable policy controversies*. New York, NY: BasicBooks.

Schwarz, M., & Thompson, M. (1990). *Divided we stand: Redefining politics, technology and social choice*. Philadelphia: University of Pennsylvania Press.

Sonneveld, M.P.W., Schröder, J.J., De Vos, J.A., Monteny, G.J., Mosquera, J., Hol, J.M.G., et al. (2008). A whole-farm strategy to reduce environmental impacts of nitrogen. *Journal of Environmental Quality, 37*, 186–195.

Stuiver, M. (2006). Highlighting the retro side of innovation and its potential for regime change in agriculture research. *Rural Sociology and Development, 12*, 147–173.

Stuiver, M. (2008). *Regime change and storylines: A sociological analysis of manure practices in contemporary Dutch dairy farming*. Dissertation. Wageningen: Wageningen UR.

Van Apeldoorn, D.F., Kok, K., Sonneveld, M.P.W.S., & Veldkamp, A. (Submitted). Panarchy rules: Rethinking resilience of agro-ecosystems.

Van der Helm, R. (2009). The vision phenomenon: Towards a theoretical underpinning of visions of the future and the process of envisioning. *Futures, 41*, 96–104.

Veldkamp, A., Altvorst, van A.C., Eweg, R., Jacobsen, E., Kleef, van A., Latesteijn, van H., Mager, S., Mommaas, H., Smeets, P.J.A.M., Spaans, L., & Trijp, van J.C.M. (2008). Triggering transitions towards sustainable development of the Dutch agricultural sector: TransForum's approach. *Agronomy for Sustainable Development, 29*, 87–96.

Vervoort, J., Kok, K., Van Lammeren, R., & Veldkamp, A. (2010). Thresholds to the future: Exploring the potential of interactive media for participatory scenarios on socio-ecological systems. *Futures, 40*, 604–616.

Chapter 5. Mobilizing consumer demand for sustainable development

Hans C. M. van Trijp[1] and Arnout R. H. Fischer[2]

[1]Hans.vanTrijp@wur.nl; TransForum and Wageningen University – Marketing and Consumer Behaviour Group; Hollandseweg 1, 6706 KN Wageningen, The Netherlands

[2]Arnout.Fischer@wur.nl; TransForum and Wageningen University – Marketing and Consumer Behaviour Group; Hollandseweg 1, 6706 KN Wageningen, The Netherlands

Abstract A lot of innovation effort is aimed at increased sustainable consumption, while at the same time actual sustainable consumption is not meeting the expectations raised by the positive public attitudes towards sustainability. This is indicative of a gap between attitudes and behaviors in sustainable consumption explained by social dilemma theory, value orientations, construal level and multiple selves theories. To bridge this gap, the social marketing approach utilizes techniques from commercial marketing to promote sustainability. Social marketing of sustainability is based on the voluntary increase in consumer motivation, opportunity and ability towards behaving sustainably. This chapter reviews evidence from ongoing TransForum research that has adopted this approach and concludes that mobilizing consumer demand for sustainable product is an essential prerequisite for growth in sustainable consumption. It also concludes that commitment of all partners in sustainable marketing chains is required to fulfill the potential of social marketing approaches in mobilizing sustainable consumption.

5.1. Introduction

Innovations aimed at sustainable development are more likely to be successful if they add value at identified places in the supply chain, in social (people), environmental (planet) and/or economic (profit) terms. In some situations the people, planet and profit values may be fully alignable, as improvements in the social and/or environmental dimensions may reduce production and/or marketing costs.[1] In such situations, moving towards a more sustainable solution does not require changes in consumer behavior. In most situations it is, however, likely that there is a complex trade-off between the three sources of value and a redistribution of associated costs and benefits across the different actors within the chain. As

---

[1] Energy saving greenhouses may be an example here that are both economically, environmentally and socially (reduced heat in the greenhouse for the workers, less light pollution at night to the neighbors) beneficial to the grower (for more in depth discussion of this invention see Chapter 2).

H.C. van Latesteijn and K. Andeweg (eds.), *The TransForum Model: Transforming Agro Innovation Toward Sustainable Development*, DOI 10.1007/978-90-481-9781-1_5,

most systemic innovations imply substantial deviations from the existing (short term profit optimized) production and marketing rationale, it is likely that, at least initially, moving towards an improved people-planet-profit balance will increase complexity within the existing supply chain and hence will be accompanied by increased costs for production and/or coordination. In those instances it can be expected that differentiation and/or focus are more promising strategic orientations of the value chain (Porter 1985).

Studies on successful innovation (see Henard and Szymanski 2001) show that the success of innovation depends on a complex set of factors related to the innovation process, corporate strategy, product-related factors and market conditions. The key factor in innovation success is the extent to which the new (and more sustainable) product or service delivers added value to the end user, in a format that is discernable and meaningful to that end user. All innovation strategies require a focus on the end user as that is the level within the value chain where the necessary (economic) value materializes in terms of 'willingness to buy' and potentially 'willingness to pay'. Thus, we assume that effective consumer demand can serve as the oil in the machinery to facilitate sustainable development. This confirms the idea of successful innovation finding its basis in sustainable competitive advantage in the market place. A focus on the (final) market as evidenced in market orientation is an important determinant of such innovation success (e.g. Kirca et al. 2005). A common definition of market orientation is the organization wide generation and dissemination of intelligence pertaining to customers and competitors, and action taken on the basis of the intelligence (Kohli and Jaworski 1990; Narver and Slater 1990).

Regarding market orientation an important distinction needs to be made between pro-active and re-active forms of market orientation (Slater and Narver 1998). Reactive market orientation strategies are customer-led in that they are responsive in nature with a focus on expressed consumer wants. Such strategies are successful and can largely be grounded in traditional consumer research methodology. However, their success is typically short-lived because consumer needs and customer satisfaction are subject to change over time. Reactive strategies focusing on adaptive changes ('incremental innovations') are necessary for the short term corporate success. Increasingly, however, it is recognized that customer-led businesses, by their very strategic focus, may actually be constrained in establishing radical or transformational innovations necessary at the systemic level. As Hamel and Prahalad (1994) put it 'the tyranny of the served market' may blind the long-term perspective. Proactive market orientation, on the other hand, is market-oriented rather than consumer led. Such businesses are committed to understanding both expressed and latent needs of their customers (Slater and Narver 1998). Market-oriented strategies are proactive in that they focus much more on customer value rather than immediate customer satisfaction. In terms of customer intelligence, proactive strategies require a focus on deep understanding of consumer behavior with an emphasis on 'what could be' rather than 'what is' and a focus on prediction rather than description. Much of the traditional market research methodology falls short in providing guidance for proactive strategies. Proactive market orientation finds its basis in more fundamental understanding of consumer behavior and uses more advanced consumer and market research methodologies such as customer observation (Leonard-Barton 1995), lead user approaches (Von Hippel 1986) and other innovative ways of user-producer interactions (Smits and Den Hertog 2007). A generative rather than adaptive learning mode is adopted, which has important implications for the type of consumer and market input that they build their strategic decisions on.

What proactive and reactive strategies have in common is their focus on the end consumer (market orientation) as a means to unlock latent demand for sustainable development. Where they differ is in their tools. Proactive market orientation strategies involve much more risk and uncertainty as they deviate from a business as usual strategy. To mitigate that risk, at this moment large scale companies regularly outsource this activity to small scale spin-off companies that can carry the risk. After proven success such initiatives are absorbed back into the mainstream company strategy.

The distinction between proactive and reactive market orientation strategies is particularly relevant in the area of sustainable development, as substantial system innovations will be required in order to meet the sustainability demands facing the world. In light of the latest projections (e.g. United Nations 2007), that by the year 2050 the world population will have grown to over 8 billion people of whom more than 57% will live in urban areas, it is evident that the current level of human consumption will not be tenable within the current production and marketing systems.

Refocusing towards sustainable development requires a delicate balance between reactive ('doing things better') and proactive ('doing better things') effort. This is a shared responsibility between multiple stakeholders: governments, consumer organization, businesses and knowledge institutions alike. Whereas in the past, environmental and social problems were primarily blamed on production methods and population growth, more recently consumption has also been explicitly identified as an important cause of these problems (Cohen 2001). If consumption is part of the problem, it seems only logical to assume that changes in current consumption patterns also constitute part of the solution toward more sustainable development. Individual social responsibility, in which consumers in word *and* action, take up responsibility for sustainable development is thus a crucial element of this part of the solution. At the moment, consumers are increasingly aware of, and concerned about the negative consequences of current consumption patterns. They accept that overconsumption may damage the systems of social and natural resources on which the consumption eventually relies. They are encouraged by governments and non-governmental organizations (NGOs) to express their concerns in opinion polls (word) but increasingly also in purchase and consumption decisions (action). Governments encourage companies to take account of sustainability in their product and marketing decisions. Companies, confronted with governmental and NGO pressures as well as more demanding consumers' pressures to take account of sustainability in their production and marketing decisions, have started to facilitate the expression of consumer concerns in actual demand, through the adoption of Corporate Social Responsibility (CSR) programs aimed at providing broader value for money.

However, despite the increasing awareness of sustainability issues among all stakeholders, one of the crucial wicked problems in developing value for sustainability and capturing that in business models lies in the fact that consumer *awareness* seems to only weakly translate into actual consumer *demand*, in terms of willingness to buy and willingness to pay for sustainable products and services (e.g. Kollmuss and Agyeman 2002; Vermeir and Verbeke 2006). From a proactive market orientation point of view a fundamental understanding of the factors underlying this discrepancy between what consumers say that matters to them and what they actually do (labeled the attitude behavior gap), is instrumental in identifying intervention strategies to close that gap. Such intervention strategies would help to promote Individual Social Responsibility through which consumers become active partners in supporting sustainable development. The resulting actual (rather than latent) demand would provide a strong basis for the multi-stakeholder development of 3P values (people, planet and profit) and 3P value capturing models necessary for system innovations with regards to sustainable development. In the long run sustainable development through transformational system innovations will only be enduring, if it is supported by actual market demand that would allow it to be deeply engrained in the production and marketing systems of companies in an economically feasible way (e.g. Meulenberg et al. 2009). From a market orientation point of view, this boils down to the key question: how to facilitate the latent demand for sustainable development into an actualized demand on which shared value creation models can be built?

This chapter shows fundamental and scientific contributions for mobilizing consumer demand for sustainable development at the conceptual and methodological level, with applications in the field of consumer demand for sustainable development. Together they focus on three principal routes for mobilizing demand: enhancing consumer motivation (e.g. through sustainable value activation and social norms), strengthening consumer ability (through information and action perspectives), and facilitating consumer opportunity (through ensuring availability and accessibility of sustainable supply).

The aim of this chapter is to share the approach chosen to close the attitude-behavior gap through motivation, strengthening ability and providing opportunity of consumers, leading to extracted key learnings obtained so far. Section 5.2, discusses the theoretical views on why the attitude-behavior gap occurs from a number of leading research streams. Section 5.3, presents the Social Marketing approach on motivation, ability and opportunity, towards the mobilization of consumer demand for sustainable development. Section 5.4, reviews key findings from ongoing TransForum research supporting this approach, structured around the three dimensions of motivation, ability and opportunity. Section 5.5 summarizes the key lessons learned, with overall conclusions provided in Section 5.6.

5.2. Attitude-behavior gaps

What consumers think and say about sustainability issues covers a broad range of opinions and attitudes, which in turn may serve a number of different 'functions' in relation to different types of behaviors. This renders the definition of attitude-behavior gaps somewhat imprecise. Attitudes are defined relative to an attitude object. Consumers may hold attitudes towards a production practice (e.g. intensive animal production), an issue (e.g. animal welfare), an object (e.g. animal friendly produced products) or a behavior (e.g. buying animal friendly produced products). Individuals have multiple options to bring their attitudes to expression (in behavior or otherwise). For example, consumers may opt for indirect influence by expressing their sustainability concerns though political votes and/or memberships of NGOs rather than translating it into individual social responsibility in their purchases and consumption.

In this chapter we specifically focus on the attitude-behavior gap in terms of individual social responsibility, that is: the discrepancy between consumers' expressed concern regarding sustainability and even their self-reported importance to express this in their individual choice, compared to the lacking occurrence of actually doing so in actual choice behavior.

In this context, the discrepancy between what consumers say (in terms of the importance of sustainable development) and what they do (in terms of their actual purchase and consumption behaviors to support such sustainable development) poses a serious challenge to actually realize a more sustainable development.

5.2.1. Limitation to the theory of planned behavior

(Positive) attitude-behavior relationships constitute a central assumption in many consumer behavior theories and *the theory of planned behavior* (e.g. Ajzen 1991) in particular. The theory of planned behavior has received a lot of attention in explaining consumer behavior since its introduction (Armitage and Conner 2000). The theory of planned behavior briefly assumes that (1) self-reported behavioral intentions are the proximal cause for actual behavior (see Schwenk and Möser 2009; Webb and Sheeran 2006), and (2) that intentions are determined by three classes of behavioral beliefs: (a) behavioral beliefs about the likely outcomes of the behavior and the evaluation of these outcomes (attitude), (b) normative beliefs about the normative expectations of others and motivation to comply with these expectations (social norm), and (c) control beliefs about the presence of factors that may facilitate of impede performance of the behavior and the perceived power of these factors (perceived behavioral control).

However, the extent to which these assumptions, and especially the relation between intention and behavior, are met is a highly debated issue. In general the combination of attitude, social norms, and perceived behavioral control leads to a reasonable prediction of behavioral intention. However, Schwenk and Möser (2009) conclude on the basis of meta-analysis in the domain of environmental behavior that:

> "Although the inferred correlation [between intention and behavior] is substantial, one should not expect to be able to explain a third of the variation observable in environmental behavior. This indicates that the strong causal claim of the Ajzen/Fishbein Model is not met in reality, calling for alternative explanations" (Schwenk and Möser 2009: 754).

This intention-behavior relationship appears to be particularly weak in the case of attitudes towards generic (abstract) issues, such as generic pro-environmental attitudes relevant to sustainable purchase (Kaiser et al. 1999). This is not unique to pro-environmental behavior, as within other rather abstract domains, such as that of health behaviors a similar conclusion has been reached, namely that traditional attitude models are not very adequate in explaining socially responsible behavior (Baranowski et al. 2003).

Several suggestions have been put forward within attitude theory to understand, explain and reduce the attitude-behavior gap. A frequently followed approach is the extension of the theory of planned behavior with additional explanatory constructs. Among those, the concept of habit strength is the most prominent and has been shown to increase the explanatory power of attitude models (Bamberg and Schmidt 2003), although it is argued that habits have little to do with planned behavior. Habits are said to activate automatic behavior, leaving nothing to explain through the theory of planned behavior (Aarts and Dijksterhuis 2000). This confirms that consumers have a status quo bias and that deviation from existing habits is a process that requires substantial effort. Another approach to amend the theory of planned behavior adopts the 'principle of correspondence'. This line of reasoning holds that attitude constructs only have explanatory power for behavior if the attitude construct is measured at the same level of specificity as the behavior it is intended to predict (see e.g. Kaiser et al. 1999). General attitudes do predict generalized behavioral patterns rather than specific behaviors and vice versa. Correspondence is defined in terms the extent to which the attitude and the corresponding behavior are measured at the same level of abstraction regarding the action involved, the target at which the action is directed, the context in which it occurs and the time of its occurrence. In other words, behaviors need to be contextualized in a relevant social and temporal perspective.

5.2.2. Socio-temporal dilemmas in choice

Social dilemma theory builds strongly on conflicting social and temporal dimensions of attitude-behavior correspondence. A social dilemma is a situation in which the interests of individuals are at odds with collective (societal) interests (Messick and Brewer 1983). Social dilemma theory largely originates from social interaction research with a strong basis in experimental game theory (such as the prisoner's dilemma game). In recent years the scope of social dilemma research has been extended to real life applications, particularly in the area of pro-environmental behavior (e.g. Van Lange and Joireman 2008). It is increasingly emphasized that social (individual versus collective) and temporal (short-term versus long-term) conflicts underlie social dilemmas (Joireman et al. 2004; Van Lange and Joireman 2008).

The *social conflict* in the context of sustainable behavior involves the fact that what is beneficial for the individual (in terms of personal need satisfaction) is not necessarily beneficial for society at large (e.g. social equality and global environmental impact) and vice versa. In many situations individuals may be required to trade off some of the personal interests for the sake of collective benefit.

The *temporal conflict* recognizes that what is satisfying in the short-term may incur costs in the future and vice versa. In the context of sustainable development this implies that consumers may have to trade-off present need satisfaction against future benefit. Inter-temporal conflicts have a long research history showing convincing evidence that compared to immediate costs and benefits, delayed consequences of behavior are undervalued in actual choice decisions (temporal discontinuity). Recent work at the interface between economics and neurosciences (e.g. McClure et al. 2007) has suggested that immediate versus delayed costs and benefits may actually be driven through different brain systems with the 'old brain', based on meso-limbic dopamine reward systems operating primarily in an affective and very impatient mode, whereas the 'newer brain', based on activity in the front parietal regions takes a more analytical approach with more room for patience for delayed consequences (LeDoux 1996). In the context of sustainable development, the temporal conflict involved between short-term gratification and longer term consideration would involve a competition between the older and the newer brain regions.

The extent to which social and temporally delayed consequences are taken into account depends on the consumer's personality characteristic *social value orientation* (the importance attached to own vs. others' outcomes, Messick 1968) and on the individual's *consideration of future consequences* (the importance attached to immediate vs. delayed consequences of one's action, Strathman et al. 1994). Extensions of social dilemma theory also incorporate Schwartz' (1977) *norm activation theory*, arguing that social and temporal orientation only materialize in actual choice behavior if (a) consumers are aware of the negative consequences of their behavior (awareness of consequences), and (b) take responsibility for mitigating the negative consequences of that behavior (ascription of responsibility).

5.2.3. Construal level theory

Social and temporal distances are also key to *construal level theory* (Liberman et al. 2007) as an explanation for attitude-behavior gaps. Construal level theory argues that how consumers go about decisions depends to a large extent on how they psychologically construe the event or decision at hand. Construal level theory places an important distinction between high level (abstract) and low level (concrete) construals. In high level construal the objects, events, or individuals are classified or represented as abstract, intangible, unobservable, broad concepts. In contrast, at low level of construal the objects, events, or individuals are represented with concrete, specific, observable, or discrete features. High (versus low) levels of construal are more likely to occur for events, objects and individuals at higher (versus lower) psychological distance, determined by the psychological framing of the decision context in terms of temporal, social and spatial distance as well as the level of hypotheticality (see Trope et al. 2007 for an extensive review).

In the context of sustainable development, asking consumers abstract questions about their opinions on sustainable development and sustainable product/service options would put them in a high level construal situation. In those circumstances of high psychological distances, consumers are more likely to base their judgments on abstract criteria and particularly on so-called ‘desirabilities’, including broad moral principles (Eyal et al. 2008). However, when confronted with actual choice decisions, where the psychological distance is low, they are likely to base their decisions on a set of concrete, specific, observable, or discrete features of the decision context (‘feasibilities’). In other words, when desirabilities and feasibilities are incongruent (as in social dilemmas), construal level theory would predict preference reversals depending on the active level of construal. Answers to abstract questions (such as general attitudes) will primarily be driven by the desirable sustainability attributes, whereas concrete action (such as choice) would be driven by a different set of considerations, namely the practical feasibilities of the choice context.

5.2.4. Multiple selves

Research in sociology, political sciences and social psychology takes the distinction between desirabilities and feasibilities in consumer context one step further, arguing that individuals are driven by two sets of considerations: the ‘what you should do’ and ‘what you want to do’. *Multiple selves theories* work from the assumption of an intrapersonal conflict where individuals are equipped with multiple selves each with different and often conflicting preferences (Read and Roelofsma 1999), and where each of these multiple selves may be triggered by different contexts. The two selves are sometimes referred to as the *want self*, fighting for whatever will bring more short term pleasure and a *should self* representing an individual’s long terms interests (Milkman et al. 2008). Frequently, and particularly in political sciences and sociology the metaphor of the consumer (want) versus citizen (should) is adopted, or the metaphor of the angel (should) and the demon (want) that coexist within the individual and strive to be heard and be acted upon, leading to decision conflict. These theoretical accounts have in common that in choice situations our inner demon seems to be a stronger guide in pushing towards short-sighted decision making rather than serving the long term interests for ourselves and society at large (Milkman et al. 2008).

In summary, a number of theories in social sciences converge on the idea that the attitude-behavior gap can be understood by the different characteristics of attitude ('should' oriented, long term, and abstract) compared to behavior ('want' oriented, here and now and concrete). Whether consumers can reasonably be expected to act upon their good intentions, depends on a number of factors related to the person (e.g. social value orientation, consideration of future consequences, awareness of consequences, ascription of responsibility), the product context (how the sustainable offering is positioned in the market place, and the size of the (perceived) social and temporal trade-offs involved) and the choice context (stimulating either the should or the want selves). The distinction between *want* and *should* is important because they highlight the difference between *why* and *how* of consumption. Want-goals drive the *why* of consumer behavior, in terms of the primary goals consumers pursue with a consumption act. Should-goals drive the direction *how* of consumer behavior in terms of the secondary benefits pursued. This is important from a sustainability perspective as consumers primarily buy food products for the sensory, social, convenience and nutritional benefits they provide, not primarily to 'save the world' (Huffman et al. 2000). However, depending on the situation, provided by e.g. the retailer providing a broad assortment, secondary benefits such as sustainability may count to move specific product choices into a sustainable direction.

5.3. Stimulating sustainable consumer behavior

Active consumer demand for sustainable development can be an important impulse for the enhancement for sustainable development that contributes to social and environmental development, yet in an economically feasible way. *Social marketing* is suggested as a relevant framework to mobilize the relevant consumer demand.

Social marketing is a relatively young discipline (Andreasen 2002) the origin of which is often traced back to an early paper by a social psychologist Wiebe who asked the question: *'Why can't you sell brotherhood and rational thinking like you sell soap?'* (Wiebe 1951/52: 679).[2]

After emerging from its early days, the field is now converging on more widely accepted definitions which can be summarized as: social marketing is the adaptation and adoption of commercial marketing activities, institutions and processes as a means to induce behavioral change in a targeted audience on a temporary or permanent basis to achieve a social goal (Dann 2010). In other words, social

[2] At that moment rational thinking was still considered the pinnacle of human decision making; to place it in perspective it is interesting to note that the influential idea of 'bounded rationality' doubting the classical approach towards rationality was coined only 5 years later (see: Simon 1957).

marketing is a sub-discipline within marketing focusing on inducing social change and effect rather than narrowly focusing on the financial objectives of the firm. It uses many of the concepts and techniques from commercial marketing, in particular the concept of voluntary exchange on the basis of perceived benefit (Houston and Gassenheimer 1987), competition (between the desired and actual behavior), and segmentation, targeting and positioning (Andreasen 2002).

Hastings (2007) suggests three crucial questions involved in the use of social marketing: (1) where are people in relation to a particular behavior, (2) what factors cause this position, and (3) how can they be moved in the desired direction? At a more strategic level the crucial challenge from a social marketing perspective is then to:

1. Understand in depth why in terms of perceived costs and benefits different audiences (segments) value current behaviors over the behaviors that are preferred from the social marketing objective (i.e. segmentation)
2. Select promising segments where improvement c.an be achieved (i.e. targeting)
3. Based on these analyses, make the needed adjustment to the offering (in terms of its price, product, and availability) and promote the exchange on the basis of perceived value (positioning)

Social marketing is one of the instruments to induce social change, besides law and education. Social marketing attempts to achieve social change by making adjustments to the current offering to enhance voluntary exchange within the competitive context. Rothschild (1999) argues that the appropriateness of social marketing, education and law depends on the level of motivation, ability and opportunity (MOA) in the target group. *Motivation* refers to a person's willingness to engage in the behavior; generally individuals are motivated to engage in a behavior when they discern their self-interest will be served by doing so. *Ability* refers to a person's skills or proficiency (e.g. self-efficacy) at solving problems including breaking a well formed or addictive habit or countering the arguments of peers. *Opportunity* is dependent on the person's environment. Lack of opportunity includes situations in which the individual wants to act but is unable to do so; for example because there is no shop offering organic produce in the neighborhood (lack of availability) or because the assortment is not well communicated or prohibitively expensive (lack of accessibility).

These applications of law, education and social marketing to promote socially desirable behavior can be plotted in a motivation, opportunity, ability grid (Table 5.1).

If the target audience is prone to behave as a result of adequate motivation, ability and opportunity, educational approaches will do in realizing the behavior change. If the condition of ability is satisfied and there is a basic motivation, but the opportunity is the limiting factor, social marketing approaches are particularly appropriate in further aligning the offering to the individual's value perception. If however, motivation is lacking whereas the conditions of ability and opportunity are satisfied there is not much that education and marketing can contribute (as both rely on free choice) and law may be the most appropriate mechanism. In all other situations a mix of approaches will be required, with education particularly suited to overcome lack of ability and law particularly appropriate to overcome lack of motivation. Marketing plays important roles in eliciting high motivation by offering more attractive opportunity that is at lower perceived costs and with greater perceived benefit.

Table 5.1. Applications of education, marketing and law depending on end user motivation, opportunity and ability (from Rothschild 1999).

Motivation		Yes		No	
Opportunity		Yes	No	Yes	No
Ability	Yes	1. Prone to behave *Education*	2. Unable to behave *Marketing*	3. Resistant to behave *Law*	4. Resistant to behave *Marketing, Law*
	No	5. Unable to behave *Education, marketing*	6. Unable to behave *Education, Marketing*	7. Resistant to behave *Education, Marketing, Law*	8. Resistant to behave *Education, Marketing, Law*

The motivation-opportunity-ability framework, although defined in rather abstract terms provides a powerful framework for structuring knowledge on mobilizing consumer demand for sustainable products and services. Figure 5.1 illustrates this framework.

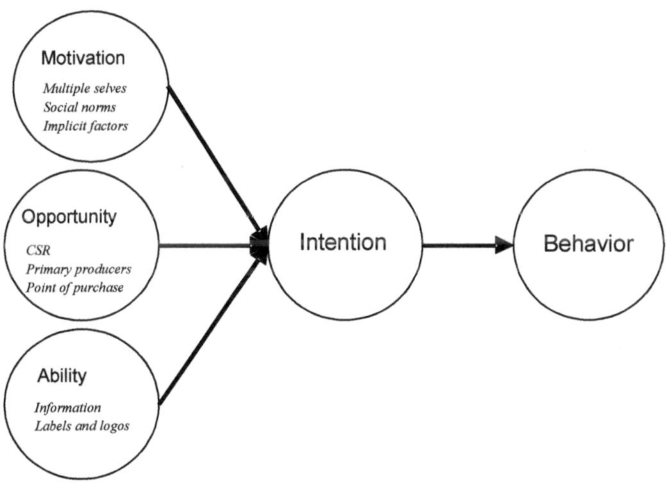

Fig. 5.1. Social marketing framework aimed at promoting sustainable consumption (italic texts refer to cases in Section 5.4).

5.4. Facilitating the three routes of mobilizing sustainable consumption

5.4.1. Route 1: Facilitating motivation

Facilitation of consumer motivation for sustainable development can progress along different routes. Most models on consumer demand, as well as on social marketing, argue that consumer motivation to engage in a (more desirable) behavior can be increased by shifting the decisional balance (i.e. the balance between personal pros and cons associated with engaging in the sustainable behavior) or by exploiting social factors in the environment through mild forms of social pressure. Several approaches to facilitate motivation to buy sustainable options have been explored.

5.4.1.1. The multiple selves in sustainable behavior

Motivation for sustainable products and services is complicated by the fact that many of the *benefits* delivered by sustainable products and services (1) manifest themselves in the long run only (rather than delivering instant satisfaction) and (2) that the benefits of sustainable behavior often accrue to others (e.g. in animal welfare) and/or society at large (e.g. in global warming). On the other hands, the *costs* involved in behavioral change, both in terms of monetary sacrifices and satisfaction of the urgent wants, accrue directly to the individual. Conceptually this is at the heart of a social dilemma as discussed before. Several lines of research have shown that the short term benefits in relation to the individual's present need satisfaction and the longer term social benefits in relation to the individual's social responsibility are motivated through different systems that compete in expressed attitudes and intentions versus real choice behavior: the so called *want* and *should* systems. These research lines converge on the finding that in more psychologically distant judgments (e.g. attitude statements, voting etc.) the *should self* tends to take priority, but that actual choices are largely dominated by the *want self.* This is a potential explanation of the attitude-behavior gap. Indeed it was shown that sustainability attributes are more likely to be important in judgment (how important in your choice), but less so when it comes to decisiveness in actual choices (determinance) (Van Dam and Van Trijp 2009a). The question is how these insights can be implemented to enhance motivation for sustainable behavior.

5.4.1.2. Social norms in sustainable development

Sustainable choices are those that bring benefits to society at large, not just now but also in the future. One way of enhancing motivation for sustainable behavior is by exploiting social factors in the environment through mild forms of social pressure. The idea is that the environment communicates social norms and that consumers hold a motivation to comply to that social pressure. In daily practice this is institutionalized through what is known as 'social norms'.

Social norms are 'rules and standards that are understood by members of a group, and that guide and/or constrain social behavior without the force of laws' (Cialdini and Trost 1998, p. 152). Social norms are thus informal, socially shared, and relatively stable guides of behavior or attitude. Their informal, non-obligatory, character implies the presence of social reinforcements, such as approval or disapproval, which distinguishes social norms from laws (Posner 1997). Additionally, social norms are shared within a group, which differentiates them from personal norms based on a consumer's own internalized values or expectations for behavior and ensures that they are generally stable over time (Jones 2006).

Social norms can take two different forms, which essentially relate to 'leading by prescription' versus 'leading by example'. Injunctive norms ('people ought to') prescribe a behavior, and refer to what people *should* do in a given situation (e.g. 'You should consume organic products'). Descriptive norms ('most people do') describe the typical behavior of others, which provides 'social proof' of what is likely to be effective behavior and sets behavioral standards from which people do not want to deviate (Schultz et al. 2007).

A meta-analysis of 200 previous studies on the effects of social norms on attitudes and intentions toward sustainable consumption (Melnyk et al. 2008) showed that (a) compared to other behavioral domains, social norms have a particularly strong effect on behavior (but not attitudes) in the field of social responsibility behavior, including sustainability issues, (b) descriptive norms (what most people do) have a stronger influence on behavior than injunctive norms (what others want you to do), whereas injunctive norms have a stronger influence on attitudes than descriptive norms, and (c) that the effects of social norms are stronger when norms are related to close and concrete sources and when the behavior at stake is public rather than private.

Differential effects of injunctive vs. descriptive norms show systematic differences in these norms. Having less cognitive resources available to deliberate on the norm (manipulated through cognitive load) limits the influence of each type of norm on attitudes and intentions, but active cognitive deliberation ('thinking more') relatively increases the effect of descriptive norms, but decreases the effect of injunctive norms on attitudes and intentions and that this effect is mediated by the amount of positive and negative thoughts that the norm formulations stimulate (Melnyk et al. 2009a).

Additionally it was shown that descriptive norms have a larger effect on intention under a condition where people strive to achieve goals, while descriptive norms are relatively less important if individuals try to avoid negative consequences of behavior (Melnyk et al. 2009c). This is even stronger for less frequent buyers of Fair Trade Coffee (the application area of the study). In contrast, the effect of injunctive norms does not show the hypothesized dependency on regulatory focus.

5.4.1.3. Implicit factors in behavior aimed at sustainable development

A major complication in the motivation for behavior aimed at sustainable development lies in the fact that many of the benefits from this behavior are uncertain to the individual in that they cannot be verified from personal experience. These are called credence qualities (Darby and Karni 1973). Individuals have two basic mechanisms at hand to form a personal belief about the product's performance on sustainability attributes (Ajzen and Fishbein 1977). The first is *informational belief formation* where the individual simply accepts relevant information from others (e.g. product recommendations from friends, consumer organizations etc.) and uses that information to form an impression on the product's sustainability. The second mechanism is *inferential belief formation*, where the individual constructs rules of thumb from available product-related information to form an impression. Inferential belief formation is a sometimes unpredictable process where the consumer makes inferences on the basis of subjective knowledge, often on the basis of subtle cues (e.g. infer sustainability from the appearance of a potato). Many of these inferences may be implicit in nature and be based on subtle cues in the product, its packaging, its communication, or its purchase and consumption context. As a result, a key question is 'to what extent can we use these subtle cues in the product and purchase context to trigger motivations of sustainable product choices'.

One option might be to cross relate socio-temporal attributes that are relevant at the moment of consumption (ego related, immediate effects) to long term sustainable goals (e.g. 'by doing this I will reduce global warming in the future'). However, activated attitudes lead most frequently to the promoted behavior if there is a fit between the temporal and social dimensions, i.e., if a sustainable action taken by me, now, leads to immediate consequences that are in proportion to my own modest effort; sustainable behavior is supported. Sustainable behavior is also supported, if the effects are framed as the result of long term group effort (a large group of people engages in this behavior leading to large effects in a more distant future). Sustainable behavior is not improved for misfits in combinations between social and temporal distances; e.g., a small action from me; leading to a large future change; or significant effort by a large group of people; leading to a small effect tomorrow (Meijers and Stapel, submitted). There is however a caveat at making the focus on close social and temporal distance (me, now) as this may specify sustainable behavior too narrowly. This may result in seemingly inconsistent behavior, for example, people fulfilling the concrete sustainability goal (e.g. purchase of organic vegetables) may reward, or compensate themselves by taking a large car to go to the shop selling these items. Therefore specific goal attribution is essential to make sure the consumers' goal is that of general sustainable behavior rather than an isolated instance of sustainable purchase (Meijers and Stapel, submitted).

5.4.1.4. The willingness to pay for sustainability labels and logo

Willingness to pay (WTP) for sustainability was studied (Van Doorn and Verhoef 2009b) through the direct WTP questioning method also used in previous studies (e.g. Cameron and James 1987; Homburg et al. 2005).

Indexed against the reference price of the product without a claim, the study reveals that willingness to pay a price premium for nutrient claims is absent, that it is modest and non-significant for fair trade claims (5%) and modest and significant for organic claims (6%). More detailed analysis reveals that claims tend to lead to negative inferences on the quality of the labeled food product (compared to the product without a claim) and that these negative quality inferences may largely account for the very limited willingness to pay for sustainable product qualities.

The very limited willingness to pay for food products with a sustainability claim found in this study relates unfavorably to what is reported in several other studies (see Van Doorn and Verhoef 2009b, for an overview). This may be (partly) due to different methodologies being adopted for the notoriously difficult challenge of measuring and predicting willingness to pay which is so crucial for the underpinning of sustainability based business models. Previous studies have sometimes reported double digit percentages WTP. Also in this particular field it seems that methodological innovation is urgently required to reach a more valid and reliable assessment of consumer willingness to pay.

5.4.2. Route 2: Facilitating ability

Because of the credence nature of sustainability attributes consumers' action perspective largely depends on information provided, either from public or private sources. The effect on actual consumer choices from a Website aimed at providing pre-competitive information on sustainability issues should confirm this. Respondents from a representative sample were selectively invited to visit an experimental website and the actual effect (if any) of the information provided on actual household purchase behavior was measured and related to respondents profile in terms of their psychographics and other background variables.

Although consumers can perceptually differentiate between the different sustainable choice motivations, when it comes to their motivational structure and priority setting they tend to lump together the sustainability considerations (e.g. environmental issues, naturalness, fair trade, local production, animal welfare, and waste) onto a single construct, differentiated from the utilitarian choice motives (taste, price, convenience), which seems in line with the construct of psychological distance (Van Dam and Van Trijp 2009b, 2009c). This is further confirmed in the analysis of the household purchase data (Van Doorn and Verhoef 2009a) which shows that also in terms of product purchases, sustainability-related products lump together on single dimensions, separate from health-related benefits and a broader category of what the authors labeled 'common and non-conscious sustainable behavior' (including semi-skimmed milk, deep litter eggs and Marine Stewardship Council labeled products). In other words, the analysis confirms at the attitudinal and behavioral level that sustainability may be a generalized concept to consumers: if you value one of the sustainability attributes, you also tend to value the other, and if you buy one type of sustainable product, you also tend to buy the other.

This raises the question what determines whether you are a sustainable consumer or household or not. Van Dam and Van Trijp (2009b) explored the predictive value of alternative theories for sustainable behavior, including value based theories, identity theory and social dilemma theory. They show that social dilemma theory explains neither self-stated importance nor decisiveness in choice. Value theory predicts self-stated importance ratings but much less so the determinance of sustainability ratings in choice. This is in line with the relevance concept as being value driven, and the determinance concept as being action driven. Identity theory explains 50% of the relevance ratings and over 30% of the determinance ratings, suggesting that of the three theories, identity theory has most potential as an explanation of sustainable consumer behavior (cf. Engelen et al. 2009).

5.4.3. Route 3: Facilitating opportunity

5.4.3.1. Corporate social responsibility activities

Given the relatively weak autonomous consumer motivation, consumer demand will need strong support from the supply side. Motivations at the supply side can come from stakeholder interactions and stakeholder group pressure, in addition to corporate social responsibility strategies. In agriculture many companies are now adopting corporate social responsibility (CSR) strategies. These strategies are being anchored into their business plan and product portfolio to guarantee long term societal license to operate, produce and sell. Different reasons for adopting sustainable strategies exist, which include added value for the final product, increased efficiency, opening new market segments, enhanced quality and quality control and improved chain management.

Research has shown, that while each argument for sustainable production on its own maybe sufficient to engage in the move towards sustainability, in reality these elements are often interlinked; you cannot improve your quality without increasing chain management for example (Ingenbleek and Frambach 2009). In a larger context, the market chain aimed at the consumer market is driven to a considerable extent by the competences and motivations of the chain captain. The influence of the chain captain affects the investments and direction of many of the other partners in the chain. Additionally, the individual chain partners influence the sustainable chain to the extent they have integrated sustainability into their business approach, and the level of innovativeness they strive for, the more innovative the company, the more it tends to favor a move towards sustainability

(Ingenbleek 2009). While this is promising for the front runners in the company population, Ingenbleek and colleagues also found that the move toward sustainability by leading companies introduces competition in the sustainability niche; and by major companies shifting their aim this also creates room for less sustainable companies to fill the gap in the cheap, large volume markets that have been left by other companies; thus ironically the move towards CSR by large companies may promote opportunistic behavior with other companies.

5.4.3.2. Enhancing opportunity from the primary production level

In the case of primary agricultural companies, transition towards more sustainable methods and approaches to production, products and services will have an important, dual effect. First of all, these products and services are essential inputs for each and any relevant sustainable agrifood chain. Secondly the importance of these stakeholders is emphasized by the fact that the farmers determine the view of the countryside, and the view consumers have of that country side. Hence in the agrifood chain, primary producers have a large and important role. Additionally, many primary agricultural companies are still small scale family operated businesses. These farms tend to be rather traditional.

Interviews have identified factors why some farms do manage to make a rather drastic transition towards sustainability in the way to operate. Results show that such transitions are supported by a specific set of farmer competences. Successfully transitioning farmers can be characterized by participation in a relevant and well formed network; by a high level of entrepreneurship; and the desire and maneuvering space to learn and apply new things (Lans et al. in prep). Entrepreneurship of primary producers in agriculture remains a topic that received little attention compared to entrepreneurship in many other sectors in the economy. Future monitoring of the development of entrepreneurship in primary agricultural production provide more in depth inside in the development of the necessary competences with farmers over the next years.

5.4.3.3. Facilitating opportunity and consumer choice at the point of purchase

The retail mix at the point of purchase plays a potentially important role in stimulating consumer demand for sustainable products, because 'as gatekeepers of the food system, supermarkets are in a powerful position to create a greener, healthier and fairer food system' (Sustainable Development Commission 2008, p. 6). To our best knowledge there exists limited systematic research into how the retail mix could be used to enhance sustainable choice behavior.

Van Nierop and Van Herpen (2010) explored the effects of retail mix on sales data in actual supermarkets. Their study shows that market share of sustainable products does not depend on the number of facings in the shelf, but in enhanced sales for sustainable products that are positioned just above the middle (bottom shelf shows strong negative effect on market share) and in the middle (rather than the far ends) of the category assortment. Grouping the sustainable products together does not have an influence on the market share of the sustainable products in our dataset, but sorting the whole product category by brand has a positive effect on market share. In contrast, sorting on price has a negative effect. Apparently, this makes it obvious at the moment of choice that sustainable products in general have higher prices than regular products.

The results further show that if the sustainable product is more expensive than the market leader, we can expect a negative effect on market share. A price cut tends to have a positive effect on the market share of the sustainable product. Market share is higher in areas with more older people and higher educated people.

By and large these results provide support for the hypothesis that supermarkets through their retail mix can enhance consumer demand for sustainable product alternatives.

In a series of follow-up smaller scale experiments (van Herpen 2009) the effect of the retail mix was studied in more depth. Together these experiments point in the direction that: sustainable product alternatives enjoy a benefit when they are structured within the assortment as a separate subcategory rather than being intermixed between non-sustainable alternatives. Although separating sustainable products is generally appreciated, it only increases the choice likelihood for sustainable products among consumers with a positive predisposition towards sustainability (i.e. their ethical orientation). However, when there are price differences between the sustainable (e.g. organic) and competitor products, consumers felt less satisfied with the separation of the organic sub-assortment and preferred the comparison products to be placed next to each other. This is most likely due to the fact that consumers may have felt uncertain of their choices due to difficulties in comparing prices.

5.4.4. Methodological challenges in stimulating consumer demand

From the previous discussion it is obvious that stimulating consumer demand for sustainable development is a multi-facetted and complex problem which can be realized from enhancing consumer demand (motivation), increasing product supply (opportunity), with knowledge and information transparency (ability) as an important facilitator. From an innovation perspective this illustrates that the development of new value propositions is a complex process. The results on low penetration rate of sustainable product alternative in the market place, suggest that food chains have only been partly successful in the mobilization of consumer demand.

Some have argued that this low success rate may at least be partly due to limitations in the prevalent model of new product development. The argument is that these traditional models have a number of limiting features:

1. Many of these models develop innovations without adequate a priori consideration of the added consumer benefit. Such innovations as a technology push have considerably higher failure rates (e.g. Cooper and Kleinschmidt 1996). User production interaction (Nahuis et al. 2009) maybe a way to introduce consumer demand in early stages of innovations, where in the context of multi-stakeholder interactions creating and managing similar visions and expectations among all stakeholders, are likely to be important tools (Zwartkruis et al. 2010). Innovativeness is a prerequisite for differentiation in the market place, but consumer meaningfulness is the determining feature for success (e.g. Szymanski et al. 2007)
2. Traditional innovation models in marketing tend to be based on consumer insight from the 'average' consumer. In other words from representative users or customers at or near the centre of the intended target market. These consumers may not be particularly helpful in informing for the radical innovation required within the system innovation approach (Lilien et al. 2002)
3. Traditional innovation models in marketing tend to select needed information from the market place only, and assign the task of generating ideas for solutions leading to new products to manufacturers (Lilien et al. 2002). This disconnects between need articulation and solution generation may seriously limited effective consumer orientation.
4. Traditional innovation models in marketing tend to assign the task of physical product development to product teams that are embedded within the current marketing and production context. This in turn may lead to a functional fixedness (myopia) where new products are more likely to stay close to existing products rather than providing the necessary breakthrough ideas (Goldenberg and Mazursky 2001). Better insights how stakeholder roles contribute to, and help overcome this barrier is a possible way out of this problem (Janzen et al. in prep).

5.5. Key lessons learned

1. *To consumers sustainability holds few dimensions.* Sustainability is a complex construct consisting of many different and potentially interacting dimensions (people, planet and profit) that may create conflicts towards sustainable development (e.g. fair trade versus food miles). When buying products, consumers tend to lump people and planet dimensions of sustainable development together in a single sustainability dimension, which they weigh against a tangible dimension (consisting of price, taste, etc.). This is confirmed both at the level of attitudes and the level of household purchases. This means that there exists something like a 'sustainable' consumer which takes account of the multiple dimensions of sustainable development at the same time.

2. *Self reported measures for positive attitudes towards sustainability are good indicators of public concern but not of market demand.* Attitudes, expressed as self-reported scores in surveys, are indicators of *should* motivations for non-tangible attributes such as sustainability. These *should* motivators (which are value driven, distant and usually abstract) are a poor indicator for actual choice (which is much more concrete, here and now, and for personal satisfaction). Despite the poor predictability of consumer attitudes towards actual behavior, the value based attitudes are an important signal for the increasing attention and concerns of consumers regarding sustainable development.
3. *Consumer demand is currently limited.* The relation between the (large) attention for sustainable development and the (modest) attention for actual demand of sustainable products is weak. The current, low market shares make data analysis of sustainable products and especially the determinants for their choice problematic.
4. *Social norms may provide a powerful tool to enhance consumer motivation for sustainability.* The leading examples of relevant peers (descriptive norms) is shown to be a potentially powerful means to enhance sustainable behavior. Leading by example seems more successful to support sustainable behavior than explicitly voiced peer pressure (injunctive norm).
5. *Active market demand is crucial for the development of viable business models for system innovation regarding sustainable development.* Companies, and supply chains more generally, can take a variety of strategies toward sustainable development. CSR can focus on improving production (less pollution, less energy use, for example) and on creating better working conditions. When CSR strategy becomes ingrained in production and marketing this requires a deviation from the currently practiced (financially optimized) business model. This requires compensation through increased market shares.

   – *Retailers can play their part, through the in-store retail mix and assortment organization.* Actual consumer demand for sustainable product alternatives can be enhanced at the point of purchase if (a) sustainable products are placed in the middle of the assortment, (b) at eye height, and (c) sustainability grouping is done within the relevant product category (but this effect is lost when super-categorical grouping on sustainable products is done, such as by creating a 'shop-in-shop').
   – *Better availability of sustainable primary products is essential.* Special competences among farmers are needed to be successful in innovation toward sustainable production. Entrepreneurship and the willingness to look outside the box are core competencies of the innovating farmer. Leading companies adopting CSR programs are introducing a standard for other companies in the same field.
   – *Sustainable demand can be enhanced through more appropriate new product development methodologies.* Early involvement of consumers is essential to build in and anchor consumer values related to use of products, social context of their use and specific attributes. Omissions of important

stakeholder groups in this phase can lead to lagging sales, protest or active rejection in a later stage.

6. *Multi-stakeholder support is crucial for sustainable innovations to succeed.* Successful innovations are the result of interactions of different stakeholders, adopting specific roles. Active stakeholder management is required to continually adjust stakeholder group configuration to changes in the project.

5.6. Conclusion

Consumer demand for sustainable production is an essential driver for the underlying sustainable production chain. However, expressive consumer support is more vocacious in the domain of opinion than in behaviors.

The social marketing perspective adopted in this chapter has shown that consumers' sustainable action can be promoted through 'nudging' the consumer towards more sustainable choice. The three routes of nudging following motivation, opportunity and ability show promise in furthering consumer choice for sustainable products. Although the results are promising, several methodological and practical issues are raised that require further investigation.

Overall the current program shows that one initiative cannot save the world, but that a comprehensive approach of small and modest steps across many stages in the product development and marketing chain can mobilize the consumption of sustainable products.

5.7. References

Aarts, H., and Dijksterhuis, A. (2000). The automatic activation of goal-directed behaviour: The case of travel habit. *Journal of Environmental Psychology*, 20(1), 75–82.

Ajzen, I. (1991). The theory of planned behavior. *Organizational Behavior and Human Decision Processes*, 50(2), 179–211.

Ajzen, I., and Fishbein, M. (1977). Attitude-behavior relations: A theoretical analysis and review of empirical research. *Psychological Bulletin*, 84, 888–918.

Andreasen, A. R. (2002). Marketing social marketing in the social change marketplace. *Journal of Public Policy and Marketing*, 21(1), 3–13.

Armitage, C. J., and Conner, M. (2000). Efficacy of the theory of planned behaviour: a meta-analytic review. *British Journal of Social Psychology*, 40, 471–499.

Bamberg, S., and Schmidt, P. (2003). Incentives, morality, or habit? Predicting students' car use for university routes with the models of Ajzen, Schwartz and Triandis. *Environment and Behavior*, 35(2), 264–285.

Baranowski, T., Cullen, K. W., Nicklas, T., Thompson, D., and Baranowski, J. (2003). Are current health Behavioral change models helpful in guiding prevention of weight gain efforts? *Obesity Research*, 11 (Suppl.), 23S–43S.

Cameron, T. A., and James, M. D. (1987). Estimating willingness to pay from survey data: An alternative pre-test-market evaluation procedure. *Journal of Marketing Research*, 24(4), 389–395.

Cialdini, R. B., and Trost, M. R. (1998). Social influence: Social norms, conformity, and compliance. In D. Gilbert, S. Fiske and G. Lindzey (Eds.), *The handbook of social psychology* (4th ed., Vol. 2, pp. 151–192). New-York: McGraw-Hill.

Cohen, S. (2001). *States of denial: Knowing about atrocities and suffering* Cambridge, UK: Polity.

Cooper, R. G., and Kleinschmidt, E. J. (1996). Winning businesses in product development: The critical success factors. *Research Technology Management*, 39(4), 18–29.

Dann, S. (2010). Redefining social marketing with contemporary commercial marketing definitions. *Journal of Business Research*, 53, 147–153.

Darby, M. R., and Karni, E. (1973). Free competition and the optimal amount of fraud. *Journal of Law and Economics*, 16(1), 67–88.

Engelen, C., Van Dam, Y. K., and Fischer, A. R. H. (2010). Because I am: Sustainable identity as a key to sustainable behaviour. submitted.

Eyal, T., Liberman, N., and Trope, Y. (2008). Judging near and distant virtue and vice. *Journal of Experimental Social Psychology*, 44(4), 1204–1209.

Goldenberg, J., and Mazursky, D. (2001). *Creativity templates in new products*. Cambridge, NJ: Cambridge University Press.

Hamel, G., and Prahalad, C. K. (1994). Competing for the future. *Harvard Business Review*, 72(4), 122–128.

Hastings, G. (2007). *Social marketing: Why should the Devil have all the best tunes?* Oxford: Butterworth Heinemann.

Henard, D. H., and Szymanski, D. M. (2001). Why some new products are more successful than others. *Journal of Marketing Research*, 37(August), 362–375.

Homburg, C., Koschate, N., and Hoyer, W. D. (2005). Do satisfied customers really pay more? A study of the relationship between customer satisfaction and willingness to pay. *Journal of Marketing*, 69(2), 84–96.

Houston, F. S., and Gassenheimer, J. B. (1987). Marketing and Exchange. *Journal of Marketing*, 51(4).

Huffman, C., Ratneshwar, S., and Mick, D. G. (2000). Consumer goal structures and goal-determination processes. In S. Ratneshwar, D. G. Mick and C. Huffman (Eds.), *The why of consumption* (pp. 9–35). New York: Routledge.

Ingenbleek, P. T. M. (2009). Resource allocation for corporate social responsibility in business-to-business markets. In prep.

Ingenbleek, P. T. M., and Frambach, R. T. (2009). Value creation from corporate social responsibility: The Role of Business model innovation. In prep.

Janzen, C., Driessen, P., and Hillebrand, B. (in prep). Theoretical framework of take off. Unpublished repot.

Joireman, J. A., Van Lange, P. M., and Van Vugt, M. (2004). Who cares about the environmental impact of cars?: Those with an eye toward the future. *Environment and Behavior*, 36(2), 187–206.

Jones, T. (2006). We always have a beer after the meeting: how norms, customs, conventions and the like explain behavior. *Philosophy of the Social Sciences*, 36(3), 251–275.

Kaiser, F. G., Woelfing, S., and Fuhrer, U. (1999). Environmental attitude and ecological behaviour. *Journal of Environmental Psychology*, 19(1), 1–19.

Kirca, A. H., Jayachandran, S., and Bearden, W. O. (2005). Market orientation: A meta-analytic review and assessment of its antecedents and impact on performance. *Journal of Marketing*, 69(2), 24–41.

Kohli, A. K., and Jaworski, B. J. (1990). Market orientation – The construct, research propositions, and managerial implications. *Journal of Marketing*, 54(2), 1–18.

Kollmuss, A., and Agyeman, J. (2002). Mind the gap: Why do people act environmentally and what are the barriers to pro-environmental behavior? *Environmental Education Research*, 8(3), 239–260.

Lans, T., Versteeg, J., and Verhees, F. (in prep). Determinants of farmer innovativeness. In prep.

LeDoux, J. E. (1996). *The emotional brain: The mysterious underpinnings of emotional life*. New York, NY: Simon and Schuster.

Leonard-Barton, D. (1995). *Wellsprings of knowledge: Building and sustaining the sources of innovation*. Boston, MA: Harvard Business School Press.

Liberman, N., Trope, Y., and Wakslak, C. (2007). Construal level theory and consumer behavior. *Journal of Consumer Psychology*, 17(2), 113–117.

Lilien, G. L., Morrison, P. D., Searls, K., Sonnack, M., and Von Hippel, E. (2002). Performance assessment of the lead user idea-generation process for new product development. *Management Science*, 48(8), 1042–1059.

McClure, S. M., Ericson, K. M., Laibson, D. I., Loewenstein, G., and Cohen, J. D. (2007). Time discounting for primary rewards. *Journal of Neuroscience*, 27(21), 5796–5804.

Meijers, M. H. C., and Stapel, D. A. (2010). Me tomorrow, the others later: How perspective fit increases sustainable behavior. Journal of Environmental Psychology, http://dx.doi.org/10.1016/j.jenvp.2010.06.002.

Melnyk, V., Van Herpen, E., Fischer, A. R. H., and Van Trijp, J. C. M. (2009a). Social norms and regulatory focus. in preparation. In prep.

Melnyk, V., Van Herpen, E., Fischer, A. R. H., and Van Trijp, J. C. M. (2009b). To think or not to think: Cognitive processes in social norms perception. The effect of cognitive elaboration on social norm influence. In prep.

Melnyk, V., Van Herpen, E. and Van Trijp, J.C.M. (2008) Social Norms as a Driving Force of Attitudes, Intentions, and Behaviour: Meta-analytical Research In: Proceedings of the 37th EMAC Conference, Marketing landscapes: A pause for thought, Brighton, United Kingdom. - Brighton, UK: University of Brighton, 37th EMAC Conference, Marketing landscapes: A pause for thought, 2008-05-27/ 2008-05-30.

Messick, D. M. (1968). Motivational basis of choice in experimental games. *Journal of Experimental Social Psychology*, 4, 1–25.

Messick, D. M., and Brewer, M. B. (1983). Solving social dilemmas: A review. *Review of Personality and Social Psychology*, 4, 11–44.

Meulenberg, M., Ingenbleek, P., and Van Trijp, J. C. M. (2009). Social responsible consumption: A general concept and its implications for marketing. Wageningen University working paper.

Milkman, K. L., Rogers, T., and Bazerman, M. H. (2008). Harnessing our inner angels and demons: What we have learned about want/should conflicts and how that knowledge can help us reduce short-sighted decision making. *Perspectives on Psychological Science*, 3, 324–338.

Nahuis, R., Moors, E., and Smits, R. (2009). User producer interaction in context: A classification, ISU working paper #09.01.

Narver, J. C., and Slater, S. F. (1990). The Effect of a market orientation on business profitability. *Journal of Marketing*, 54(4), 20–35.

Porter, M. E. (1985). *Competitive advantage*. New York: Free Press.

Posner, E. A. (1997). Standards, rules, and social norms. *Harvard Journal of Law and Public Policy*, 21(1) 101–117.

Read, D., and Roelofsma, P. (1999). Hard choices and weak wills: The theory of intrapersonal dilemmas. *Philosophical Psychology*, 12(3), 341–356.

Rothschild, M. L. (1999). Carrots, sticks, and promises: A conceptual framework for the management of public health and social issue behaviors. *Journal of Marketing*, 63(4), 24–37.

Schultz, P. W., Nolan, J. M., Cialdini, R. B., Goldstein, N. J., and Griskevicius, V. (2007). The constructive, destructive, and reconstructive power of social norms: Research article. *Psychological Science*, 18(5), 429–434.

Schwartz, S. H. (1977). Normative influences on altruism. In L. Berkowitz (Ed.), *Advances in Experimental Social Psychology* (Vol. 10). New York: Academic Press.

Schwenk, G., and Möser, G. (2009). Intention and behavior: A Bayesian meta-analysis with focus on the Ajzen–Fishbein Model in the field of environmental behavior. *Quality and Quantity*, 43(5), 743–755.

Simon, H. A. (1957). *Models of man: Social and rational*. New York: Wiley.

Slater, S. F., and Narver, J. C. (1998). Customer-led and market-oriented: Let's not confuse the two. *Strategic Management Journal*, 19(10), 1001–1006.

Smits, R. E. H. M., and den Hertog, P. (2007). TA and the management of innovation in economy and society. *International Journal of Foresight and Innovation Policy*, 3(1), 28–52.

Strathman, A., Gleicher, F., Boninger, D. S., and Edwards, C. S. (1994). The consideration of future consequences: Weighing immediate and distant outcomes of behavior. *Journal of Personality and Social Psychology*, 66(4), 742–752.

Sustainable Development Commission. (2008). Downloaded from www.sd-commission.org.uk /publications/downloads/GreenHealthyAndFair.pdf.

Szymanski, D. M., Kroff, M. W., and Troy, L. C. (2007). Innovativeness and new product success: Insights from the cumulative evidence. *Journal of the Academy of Marketing Science*, 35(1), 35–52.

Trope, Y., Liberman, N., and Wakslak, C. (2007). Construal levels and psychological distance: Effects on representation, prediction, evaluation, and behavior. *Journal of Consumer Psychology*, 17(2), 83–95.

United Nations. (2007). World urbanization prospects: The 2007 Revision Population Database (Publication. Retrieved November 16, 2009: http://esa.un.org/unup/index.html.

Van Dam, Y. K., and Van Trijp, J. C. M. (2009a). Dimensionality of sustainability. Unpublished manuscript, Wageningen.

Van Dam, Y. K., and Van Trijp, J. C. M. (2009b). Consumers and sustainable development. Effect of Meer over Eten website on consumer attitude and sustainable purchase behavior. Unpublished manuscript, Wageningen, Nl.

Van Dam, Y. K., and Van Trijp, J. C. M. (2009c). The "Meer over Eten" project: General overview and background. Unpublished manuscript, Wageningen, Nl.

Van Doorn, J., and Verhoef, P. C. (2009a). Construction and explanation of an index of sustainable purchasing behavior in the Dutch Food Market. .Unpublished manuscript, Groningen, Nl.

Van Doorn, J., and Verhoef, P. C. (2009b). Willingness to pay for organic, fair trade, and nutrient food claims: differences between utilitarian and hedonic products. Unpublished manuscript, Groningen, Nl.

Van Herpen, E. (2009). Stimulating consumer choice for sustainable products at the point of purchase. Progress Report. TransForum.

Van Lange, P. A. M., and Joireman, J. A. (2008). How we can promote behavior that serves all of us in the future. *Social Issues and Policy Review*, 2(1), 127–157.

Van Nierop, E. and Van Herpen, E., (2010), The influence of product clustering on sales of organic and light food products, Paper presented at the Marketing Science Conference, 2010, Koln.

Vermeir, I., and Verbeke, W. (2006). Sustainable food consumption: Exploring the consumer "attitude – Behavioral intention" gap. *Journal of Agricultural and Environmental Ethics*, 19(2), 169–194.

Von Hippel, E. (1986). Lead users: A source of novel product concepts. *Management Science*, 32(7), 791–805.

Webb, T. L., and Sheeran, P. (2006). Does changing behavioral intentions engender behavior change? A meta-analysis of the experimental evidence. *Psychological Bulletin*, 132(2), 249–268.

Wiebe, G. D. (1951/52). Merchandising commodities and citizenship on television. *Public Opinion Quarterly*, 16(Winter), 679–691.

Zwartkruis, J. (2010) "Interaction in Innovation Processes from a framing perspective; Collaboration in finding the golden egg." Presentation at 9th Annual meeting of the Technology, Management and Policy Graduate Consortium, 27–29 June 2010, Cambridge, United Kingdom.

Chapter 6. From motivating assumptions to a practical innovation model

H. Christopher Peterson[1] and Sander E. Mager[2]

[1]Peters17@msu.edu; Nowlin Chair of Consumer-Responsive Agriculture, Michigan State University, USA

[2]Mager@TransForum.nl; TransForum, P.O. Box 80, 2700 AB, Zoetermeer, The Netherlands

Abstract In trying to formulate a practical innovation model, a meta-analysis has been carried out testing the effectiveness of the five motivating assumptions described in Chapter 1. Experiment performance of 30 action experiments is assessed across 14 attributes related to impact and process outcomes. The experiments are grouped into quartiles of performance, and 76 explanatory attributes are examined to determine their influence on performance. The results strongly support the effectiveness of the motivating assumptions in the formation, guidance and process monitoring of the experiments. In addition to the five assumptions, two core strategies are found to be critical to experiment performance. When these findings are combined with the findings from monitoring in action, a practical innovation model emerges. Although the findings are subject to limitations, the model appears to have good prospects for general application to other sustainable development settings.

6.1. Introduction

Chapter 1 framed the 'formidable double challenge of expanding [agricultural] production and at the same time improving the way in which this production takes place'. Creating dual progress in productivity and sustainability is a wicked problem that transcends normal innovation and demands deeper system innovation. System innovation will make it possible 'to dramatically change the mode of agricultural production and devise new ways of food production that do not waste renewable resources and that do not put a heavy burden on social and environmental systems'. But how is this done? The various chapters since Chapter 1 have suggested pieces to the puzzle of how, but the pieces have not yet come together. The search continues for a holistic new innovation system.

The framework of five motivating assumptions that are applied in the action experiments give critical clues and insights into the broader system innovation model. By examining the performance and execution of the action experiments, we can explore what worked and why, as well as draw the meta-learning from across the collective experience. This learning becomes a guide to those who are

H.C. van Latesteijn and K. Andeweg (eds.), *The TransForum Model: Transforming Agro Innovation Toward Sustainable Development*, DOI 10.1007/978-90-481-9781-1_6,

also seeking to manage other action experiments within the wicked problem of sustainable development for agriculture or any similar field of endeavor. For more information on the action experiments and the TransForum program, please see Appendix I.

The five motivating assumptions are:

1. Sustainable development is a dynamic process.
   No closed form definition exists for the concept. It must be given realizable meaning in a multiple of valued dimensions that evolve across time.
2. Sustainable development needs system innovation.
   More of the same is simply not enough. The hardware, software, and orgware of agriculture must be innovated if movement toward sustainability is to be achieved.
3. System innovation is a non-linear learning process.
   The normal science approach of problem-solution-application must give way to a messier process of consensus goal setting-joint knowledge creation-reflexive learning.
4. System innovation requires a multi-stakeholder approach.
   All stakeholders bring existing knowledge and concern to the process, and their collective presence is needed for legitimacy and productive creativity.
5. Multi-stakeholder approaches imply trans-disciplinary knowledge creation. *Complex, messy problems do not know disciplinary bounds. All relevant skills and knowledge must be combined and extended to create the new knowledge that will evolve the system.*

In reflecting on the action experiments, it is acknowledged anecdotally that the action experiments had multiple levels of performance, and this result is confined in the more formal performance ratings later in this chapter. The variability of performance allows conclusions to be drawn about the efficacy of the five motivating assumptions and, at least provisionally, how generally these conclusions may be applied to managing other wicked situations.

6.1.1. Measuring and explaining performance

The meaning of performance (let alone its measurement) in the context of wicked problems is not entirely clear. Wicked problems are by definition unsolvable (Rittel and Webber 1973). After all, how does one solve something that is indefinable in the first place, morphs every time something is done about it, and has no definitive measure of attainment? Wicked problems can only be managed. Performance needs thus to be based on the outcomes of managing the thing, and not solving the thing. Two sets of outcomes would seem to matter in this context:

1. desired movement in impact outcomes toward, rather than away from, the goal of managing the wicked problem (in the case of sustainable development this can mean improved outcomes related collectively to people, planet, and profit), and;

2. desired movement in process outcomes that evidence constructive engagement of relevant stakeholders in the problem management (e.g., relevant stakeholders are at the table, effectively engaged with each other, and supportive of actions and learning achieved).

Together (1) and (2) are necessary and sufficient for performance in managing wicked problems. If you have (1) and not (2), a strategy to achieve impact exists but it may be vetoed in implementation by some disgruntled or excluded stakeholder. If you have (2) and not (1), process gains may make for a supportive environment for implementation, but no meaningful impact strategy exits to implement. With (1) *and* (2), both an implementable strategy and a supportive implementation environment exist. Actual performance then hinges on how well (1) and (2) are carried out in practice. Performance measurement thus has both an impact component and a process component.

To test the performance of the five motivating assumptions, a hypothesis is needed concerning the cause and effect relationship between the model and performance. The existing literature on wicked problems makes it clear that no one approach to managing them can work (see Regeer 2010). As with managerial problems generally, the lack of a single or universal strategy to create performance simply means that any selected strategy will be more or less effective based on the extent to which the strategy fits the context of the situation being managed. In other words, any tested strategy is effective or ineffective contingent on its fit to the context in which it is applied. The performance hypothesis for this analysis becomes:

A sustainable development action experiment will achieve high performance in impact and process outcomes if the action experiment's processes effectively implement the five motivating assumptions and fit the action experiment's internal and external contexts.

The remainder of the chapter reports on a test of this hypothesis. In doing this, the sections of the chapter are as follows: specification of the methodology, performance measurement of the action experiments and its overall implications, the influence of various explanatory attributes on action experiment's performance, reporting on related process findings from other research on the process, and overall conclusions.

6.2. Methodology

A grounded theory approach is used in this research (Strauss and Corbin 1990). The analysis starts with the performance hypothesis posed in the introduction, and uses a theory-testing-revision cycle to induce the learning that results in a more specified hypothesis for future testing. The empirical evidence is collected using case based research methods. The initial hypothesis will thus be tested on 30 action experiments from the TransForum portfolio. Each and every action experiment becomes a *case*. Combining grounded theory and case method is appropriate given the exploratory nature of this research into how sustainable development can be managed. The research needs to consider a wide variety of potential performance and explanatory attributes in the context of a limited number (30) of instrumental cases as the empirical context. As consistent with case method, 'how' and 'why' questions (Yin 2009) concerning sustainable development are of paramount importance to exploring the hypothesis.

The logic of the research analysis is as follows:

1. The level of performance is assessed for each case, based on a set of impact and process outcome attributes.
2. The cases are clustered into performance groups by like levels of performance.
3. A large set of explanatory attributes (independent variables) are created and used to assess their influence on the performance of each case.
4. The influence assessments for the explanatory attributes from each case are accumulated for each performance group.
5. The patterns of explanatory attributes are compared and contrasted within the performance groups and across the groups. The patterns that appear to have explanatory power in relation to performance form the synthesized learning from the cases. This learning is used to revise the performance hypothesis for the next round of grounded theory.

This analysis is done qualitatively based on descriptive statistics (e.g. percentages of cases influenced or in a category of analysis). The challenges of small numbers of cases (especially in various categories) and large numbers of potentially relevant explanatory attributes (creating identification concerns) preclude standard statistical analyses.

The assessments of performance and explanatory attributes are critical to the analysis. Based on the mix of information required for the analysis, the primary source of information was determined to be an extensive attribute assessment survey. Each case was examined by a panel of three jurors who have a deep understanding of the cases being assessed. These jurors individually filled out the survey for each action experiment assigned. The assessment ratings for the various attributes are combinations of these independent juror ratings. Thus, the research methodology hinges heavy on making the tacit knowledge of these deep insiders explicit and open to analysis.

The information of the attribute assessment instrument was triangulated (Yin 2009) in a number of ways. First, in-depth interviews were held to gain insights into the face validity of the results. Second, the jurors considered the performance ratings themselves and explored the specific strategies used within the cases for innovation management. And thirdly, written case reports were analyzed for confirming and disconfirming information about the performance and processes.

The jurors assessed the cases on performance, both on impact and on process. The impact of an experiment relates to its triple bottom line or 3P (people, planet & profit) sustainability outcomes and to the realization of its own goal in producing a plan, an investment in implementation or a functional new business leading to actual value creation. The process performance of an experiment relates to aspects that follow from the motivating assumptions. Eight attributes were included in the assessment survey to account for that:

1. Innovation happened (assumption 2)
2. Action learning happened (assumption 3)
3. All relevant KENGi[1] partners engaged in the process (assumption 4)
4. The KENGi partners took ownership of the project (assumption 4)
5. Collaboration among the KENGi partners happened (assumption 4)
6. The action experiment will continue after TransForum
7. New knowledge was created (assumption 5)
8. Trans-disciplinary collaboration happened (assumption 5)

Next to that, the jurors scored on 76 explanatory attributes to clarify the cause of the performance. The 76 explanatory attributes are divided in 27 internal attributes that represent aspects of the action experiment over which the experimenters have some control, 22 external attributes that represent economic, societal, environmental, political, and physical boundary conditions, and actors with which the action experiment must interact and operate but over which the action experiment has little control and 27 process attributes that represent strategies and process used to operate and control the action experiment. Appendix IV gives an overview of all attributes assessed in this research.

6.3. Performance of cases and assumptions

This section focuses on the assessment of performance for both the action experiments themselves and for the five motivating assumptions as a whole. The analysis of the explanatory attributes follows in the next section.

6.3.1. Performance groups

The performance results of the attribute assessment instrument are presented in Table 6.1. The ratings are expressed as the averages of the Likert scale assessments from the three jurors who assessed each case. This average was used given no compelling logic for weighting the individual components differently. The 30 cases are divided into four performance quartiles: the high performance quartile (including 8 cases), the moderate high quartile (7 cases), the moderate quartile (7 cases) and the low quartile (8 cases).

[1] KENGi is an acronym for the four key partners that each action experiment was required to have: (1) Knowledge institutions, (2) Entrepreneurs and business partners, (3) Non-governmental or civil society organizations, and (4) Government bodies. The small "i" indicates the goal of working together – innovation.

Table 6.1. Performance ratings by performance attribute and performance quartile.

Performance Attributes	High Quartile Average	Moderate High Quartile Average	Moderate Quartile Average	Low Quartile Average
***Impact Outcomes** (1 low-5 high)*				
People	3,5	3,0	2,4	1,4
Profit	3,3	3,1	2,6	1,7
Planet	3,9	2,7	2,5	1,7
Plan, feasibility study, or like	4,3	3,9	3,5	1,7
Investment in implementation	3,1	1,7	2,2	1,0
Harvest (actual value created)	2,0	1,4	1,8	1,1
Average impact rating	***3,6***	***3,1***	***2,8***	***1,5***
Average impact rating range	***3,3-4,2***	***2,8-3,2***	***2,1-3,1***	***1,0-2,2***
***Process Outcomes** (1 low-5 high)*				
Innovation happened	3,8	3,6	2,7	1,4
Action learning happened	3,8	3,1	2,9	1,5
All KENGi partners engaged	3,9	3,2	3,0	1,6
KENGi partners took ownership	3,6	3,3	2,5	1,0
Collaboration happened	3,9	3,4	2,9	1,4
Continues after TransForum	3,9	3,9	3,8	1,6
New knowledge created	4,2	3,5	3,2	1,8
Trans-disciplinary collaboration	4,1	3,1	2,6	1,5
Average process rating	***3,9***	***3,4***	***2,9***	***1,5***
Average process rating range	***3,7-4,4***	***3,1-4,0***	***2,8-3,2***	***1,0-2,3***
Total Performance Rating*	***7,5***	***6,5***	***5,7***	***3,0***
Total Performance Rating Range	***7,0-8,3***	***6,1-6,8***	***4,9-6,1***	***2,0-4,2***

*Total Performance Rating = Average Impact Rating + Average Process Rating; Likert scale; 2 low-10 high.

The data in Table 6.1 suggest a number of interesting findings:

- *Impact and process performance move together.* One does not observe strong impact performance linked to weak process performance or vice versa. The individual case data also support this finding. Performance on both dimensions appears to be linked. This finding is consistent with the motivating assumptions. For the cases in the top three quartiles, the average process rating is slightly better than the average impact, but it is not clear that the differences are significant.
- *3P performance is balanced across the three Ps and declines rather uniformly across the quartiles.* This pattern across the 3Ps would in fact be the preferred one as opposed to observing the dominance of one over the others. People impacts do tend to slightly lag the other two. Interview findings support that less attention was paid to the people-P in many of the cases.

- *Impact performance falls off most sharply in all quartiles for investment and harvest performance.* Most cases were focused on planning, and this fall off is expected. However, ultimate impacts of sustainable development must be based on investment and harvest performance. Timelines longer than 5 or 6 years would appear to be needed to achieve performance in these added dimensions.
- *A substantial range of performance is observed.* The jurors apparently did not engage in skewing the rating of case performance toward all cases being successful.

6.3.2. Performance of the five motivating assumptions

As a first step in testing the five motivating assumptions for their effectiveness, Table 6.2 presents a number of case descriptors and then traces the relationship of these descriptors to the performance quartiles. These descriptors arise from important issues in sustainability and in the history of the case portfolio.

- The 'When Project Was Selected' descriptor is based on the state of the five motivating assumptions when an action experiment started. Seven action experiments in the portfolio were selected prior to the existence of the full set of motivating assumptions. They are called the pre-assumptions cases in the table. Their presence represents a natural experiment within the portfolio on the effectiveness of the assumptions as a guide to selection. The distinction between early and late assumptions represents cases selected after the motivating assumptions were first put in place and after the application of the motivating assumptions evolved into their final form.
- The 'Wickedness' descriptor provides evidence that some cases were more wicked while others were less. All projects were done in the wicked arena of sustainable development, but some focused on a more technical evaluation or a less controversial aspect of a wicked problem. As will be presented shortly, this distinction provides some additional insight about when the five motivating assumptions should or should not be applied.
- The 'Potential Application' descriptor shows that some cases were designed to have their learning applied to mainstream agriculture versus applied to niche or specialty agriculture. Traditionally some advocates on both sides of the scale issue have argued that sustainability is more or less achievable based on scale. Some insight into this issue relates to this division cases.
- Finally, the 'Type of Innovation' descriptor divides the cases into several different categories based on the differing types of innovation inherent in the individual cases. Six distinct types of innovation can be indentified across the portfolio and this diversity is important in ultimately determining how general the five assumptions may be for application elsewhere.

Table 6.2 provides several performance comparisons based on these different case descriptors.

Table 6.2. Performance comparison by case descriptors.

Case Descriptor	Number of Cases	Percent in Top 3 Quartiles	Percent in High Quartile	Percent in Low Quartile
When case was Selected				
Pre-assumptions	7	43%	14%	57%
Early assumptions	15	80%	33%	20%
Late assumptions	8	88%	25%	13%
Potential Application*				
Mainstream	15	80%	27%	20%
Niche	8	88%	38%	13%
Wickedness*				
More Wicked	15	93%	40%	7%
Less Wicked	8	63%	13%	38%
Type of Innovation*				
Agropark (single-site coordination)	3	67%	33%	33%
Chain Coordination	5	80%	40%	20%
Regional Coordination	7	86%	14%	14%
Develop New Chain	2	100%	50%	0%
Revenue Innovation	2	100%	50%	0%
Cost Innovation	4	75%	25%	25%

*Descriptor distribution based on 23 cases that were selected by early and late principles. It is only coincidence that the distribution of early/late, mainstream/niche and wicked/less wicked all happen to be 15/8 by number. There is no significant correlation between the three categories.

In regard to the natural experiment on the effectiveness of the five motivating assumptions within its portfolio, case performance was substantially lower when cases were selected in the pre-assumptions phase than in either the early or late phases.[2] The late group fairs slightly better than the early group in performance, but given the small sample sizes it is hard to judge the significance of the difference. Cases selected *and* operated with the five motivating assumptions faired better performance-wise. The remainder of Table 6.2 only looks at the 23 cases selected in the early and late assumptions period, when the motivating assumptions were in place for both selection and operation of the cases.

In regard to the scale of application, the distribution of cases into high and low performance quartiles appears to show no significant difference between the mainstream and niche action experiments. Niche action experiments have a slight edge, but both scales appear to allow for well performing action experiments.

In regard to wickedness, the finding is perhaps more interesting. The more-wicked cases had higher performance outcomes with applying the five motivating

[2] 57% in the low performance quartile for the pre-assumptions group vs. 20% and 13% for the early and late groups respectively.

assumptions than did less-wicked cases. As will be shown in the explanatory attributes section, the five motivating assumptions dictate significantly more process management than a non-innovation, non-wicked problem might need. The performance finding here suggests that this extra process load may be detrimental to less wicked problems. The finding indicates that the five motivating assumptions are effective for more wicked problems while it should likely not be used in less wicked situations.

The lower panel of Table 6.2 shows the performance comparison across the six types of innovation identified. Given the small sample sizes, it is difficult to conclude anything definitively. However, cases that focus on the complex coordination problems of existing chains or single-site coordination of multiple enterprises appear to have some greater performance challenge than cases starting a new chain or focused on revenue enhancement (more under the control of a single or a limited number of entities). This distinction likely merits additional research in other case settings because type of innovation may be a significant contingency in selecting process management strategies. The evidence here shows that a number of different types of innovation achieved effective performance. This finding suggests that the five motivating assumptions may be broadly applicable to various types of innovation.

6.4. Influence of explanatory attributes and links to motivating assumptions

This section moves from the more general consideration of performance itself to the specifics related to causes of performance. As outlined in the methodology section, 76 potential explanatory attributes were used in the attribute assessment instrument. The explanatory attributes are now organized and reorganized in various patterns based on the performance quartiles. These patterns are compelling and imply potentially strong casual relationships.

Given the qualitative nature of the analysis (descriptive statistics, not inferential), the issue of when to say that an explanatory attribute has importance or potential causality is not a trivial one. Spurious correlations must always be a concern. As a convention, any attribute that affects half or more of the case in a quartile will be said to be important or have potential explanatory power. The 50% benchmark has face validity for this exploratory analysis. In practice, a factor that could be influential a third of the time may also be worth tracking as a decision maker. The reader is left to decide where the relevant benchmark actually falls given personal and professional needs.

6.4.1. Explanatory attribute categories and performance

Table 6.3 provides an overview of the influence levels of the various explanatory attribute categories.[3]

An examination of the table shows that the attribute categories for size/scope, external conditions and external actors were the least influential across the quartiles. But even here, the top two performance quartiles had approximately 1/3 of their cases positively influenced by the categories. The absence of influence by external conditions at least suggests that case performance was not driven by extremes in external conditions. In other words, high performance cases were not high performance merely because they had strongly positive external conditions and supportive external actors. Similarly, low performance cases were not low performance merely because of a strongly negative external context.

Wickedness is the only category that in balance tends to have a negative influence on cases although it might be expected to be even more negative than it is. This finding may also reflect the fact from Table 6.2 that roughly a third of the cases were less wicked in general than might have been expected.

The categories of general project traits (5 attributes), leadership (2) and general process management (7) all appear to be reasonably influential – on the positive side for the upper two quartiles and negatively for the low quartile. The finding here is that general project management practices are relevant even to innovation. In regard to the attributes included in the assessment, vision and objectives are especially influential. Strategies for communicating (internally and externally), overcoming obstacles and unexpected conditions, aligning with the case's external context, and using process phases effectively were all influential as well. Leadership (who led and how) is also strongly influential. Because these attributes of project management are already well studied in the academic and professional practice literatures, nothing more will be done in analyzing these categories here.

However, the in-depth interviews made it clear that standard project management strategies need adjusting in one critical respect for innovation – they must be applied more flexibly so as not to stifle the very innovation sought. For example, milestones and project schedules are equally critical to an innovation project as to a standard project in order to gauge and drive project progress, but they need to be reviewed and adjusted more often for process loopbacks and new directions. But this need for extra flexibility should not be interpreted as meaning that more general project management practices can be ignored or managed ineffectively. The portfolio with action experiments suggests quite the opposite.

---

[3] Although the influence levels for each of the 76 explanatory attributes will not be shown, the maximum number of cases influenced by a particular attribute was 28 of the 30 projects; the minimum number of case influenced was 3; and, the number of attributes that influenced 50% or more of the cases was 45 out of the possible 76. No one attribute could be eliminated as totally lacking in influence.

The remaining categories all show evidence of substantial influence both positive and negative on the cases. They will each be analyzed at the individual attribute level in the next subsection.

Table 6.3. Influence of attribute category and by performance quartile.

	Average Percent of Cases Positively* Influenced by Attribute Group and by Performance Quartile				Average Percent of Cases Negatively* Influenced by Attribute Group and by Performance Quartile			
Attribute Categories	**HQ**	**MHQ**	**MQ**	**LQ**	**HQ**	**MHQ**	**MQ**	**LQ**
Internal Context (27 attributes in total):								
Initiating Conditions (3)	100%	76%	62%	33%	4%	24%	24%	58%
Project Membership (4)	72%	61%	39%	16%	9%	11%	18%	50%
Wickedness (3)	42%	38%	33%	8%	38%	19%	38%	54%
Potential for Conflict (7)	43%	37%	18%	9%	7%	14%	20%	54%
General Project Traits (5)	65%	60%	34%	23%	13%	20%	23%	33%
Size and Scope (5)	33%	43%	31%	10%	20%	20%	20%	13%
External Context (22 attributes):								
External Conditions (15)	31%	29%	24%	8%	20%	10%	15%	16%
External Actors (7)	41%	31%	14%	2%	18%	16%	12%	27%
Process (27 attributes):								
Initiating Conditions (3)	63%	48%	29%	17%	4%	19%	24%	54%
Leadership (2)	63%	57%	43%	19%	19%	0%	29%	63%
General Process Mgmt. (7)	70%	55%	35%	11%	5%	8%	10%	45%
Innovation Management (11)	74%	53%	42%	19%	2%	5%	8%	30%
Impact of Intermediary (4)	56%	43%	50%	9%	6%	11%	21%	50%

*Positively and negatively influenced were determined by respectively +2 or +3 and −2 or −3 on the attribute influence scale. Quartile percentages can add to more than 100% due to the possibility of strongly conflicting ratings among the raters.

6.4.2. *Explanatory attributes, core strategies and the motivating assumptions*

Several categories of attributes are now explored at the individual level. The attributes fall into three main areas:

(1) initiating conditions, both for an action experiment and for its process;

(2) influences related to motivating assumption 4 (multi-stakeholder engagement), and;
(3) influences related to motivating assumptions 2 (system innovation), 3 (non-linear learning), and 5 (trans-disciplinary collaboration).

The level of influence is analyzed in each area.

6.4.2.1. Influence of initiating conditions

The analysis of Section 6.3.2 (see especially Table 6.2) already established that cases selected before the motivating assumptions were in place had substantially lower performance than those cases selected *and* executed under the assumptions. This finding suggests that initiating conditions surrounding the formation and selection of an action experiment have critical importance. This result is further substantiated by two categories of explanatory attributes.

Table 6.4 shows the results related to these two categories by individual attribute. Initiating case conditions are especially influential: on the positive side for the top three quartiles and on the negative side for the low quartile. Across the top three quartiles, a marked decline in percentage of cases influenced is observed. The high quartile is 100% for all three attributes; the moderate high quartile is above 70% for all three; and, the moderate quartile is at or above 57% for all three. Even the low quartile was positively influenced by the 'why' of the case initiation.

Table 6.4. The influence of initiating conditions on case performance.

	Percent of Cases Positively* Influenced by Attribute and by Performance Quartile				Percent of Cases Negatively* Influenced by Attribute and by Performance Quartile			
Attribute	**HQ**	**MHQ**	**MQ**	**LQ**	**HQ**	**MHQ**	**MQ**	**LQ**
Initiating Conditions								
How experiment was initiated	100%	71%	57%	25%	0%	57%	14%	75%
Who initiated experiment	100%	71%	57%	13%	0%	14%	43%	75%
Why experiment was initiated	100%	86%	71%	63%	13%	0%	14%	25%
Initiating Process Conditions								
How intentionally process was selected	75%	43%	29%	25%	0%	43%	43%	25%
How organically process arose as project unfolded	63%	57%	43%	13%	0%	0%	14%	63%
How process was initially managed	50%	43%	14%	13%	13%	14%	14%	75%

*Positively and negatively influenced were determined by respectively +2 or +3 and −2 or −3 on the attribute influence scale. Quartile percentages can add to more than 100% due to the possibility of strongly conflicting ratings among the raters.

Initiating process conditions were somewhat less influential as a category, but relatively high in influence nonetheless. For the high quartile, all three attributes are at 50% or more of the cases. The moderate high quartile had 43% or more of its cases positively influenced. The positive influence falls off substantially for the moderate and low quartiles, while the negative influence is strong for the low quartile in regard to how organically the process unfolded and how the process was initially managed.

Both of these category findings suggest that project and process initiation is a critical phase in the life of an action experiment.

6.4.2.2. Influence of attributes related to assumption 4 (multi-stakeholder engagement)

The motivating assumption 'system innovation requires a multi-stakeholder approach', operationally was translated in the action experiments into the requirement that the relevant so-called KENGi partners from Knowledge institutes, Entrepreneurs and business partners, Non-governmental and societal organizations, and Government bodies were present and engaged in the innovation effort.

The requirement to have the relevant stakeholders engaged is the one held assumptions of managing wicked problems more generally (Peterson 2009). One of the criteria that define wicked problems is that there is no definite formulation of the problem (Rittel and Webber 1973). Stakeholders can have deeply held and conflicting views of the problem. If stakeholders are not part of the management process, then these conflicting views would be predicted to result in stakeholders excluded from the process opposing the outcomes of the process and potentially blocking implementation of the invention or innovation.[4] This is why three of the eight process performance attributes relate to assumption 4.

In observing and rating the actual conduct of the cases, three sets of explanatory attributes were considered concerning assumption 4:

- *Project membership.* In assessing the influence of this category it was equally important to consider who was not engaged as well as who was engaged.
- *Potential for conflict.* One would expect this potential to be present in a multi-stakeholder team dealing with a wicked problem and the influence of that presence to be negative. The absence of such influence would suggest a better performing team.
- *External actors.* If the multi-stakeholder team was improperly formed in its inclusion of the relevant KENGi partners, then those not present on the team become outside or external actors who would be expected to exert negative influence on the action experiment. Absence of this negative influence would again suggest a better performing project team.

---

4 See also Chapter 4 on the role of images in the implementation of innovations.

Table 6.5 presents the influence assessments related to the various attributes on project membership, potential for conflict, and external actors. The results are broadly supportive of motivating assumption 4.

Three of the four membership attributes are strongly positive for the upper three quartiles although the percentage of influence does decline across the quartiles. The most negatively influential attribute is as expected 'who was not engaged' which had strong negative influence in the low quartile and affected more than half the cases in the moderate quartile. Recall that half of the low quartile cases were selected prior to the motivating assumptions being in place, including of course motivating assumption 4.

Table 6.5. The influence of attributes related to motivating assumption 4 (multi-stakeholder engagement) on case performance.

	Percent of Cases Positively* Influenced by Attribute and by Performance Quartile				**Percent of Cases Negatively* Influenced by Attribute and by Performance Quartile**			
Attribute	**HQ**	**MHQ**	**MQ**	**LQ**	**HQ**	**MHQ**	**MQ**	**LQ**
Project Membership								
Who engaged	100%	86%	57%	25%	0%	0%	0%	38%
Who was not engaged	13%	0%	0%	0%	25%	43%	57%	88%
How 'who' changed	88%	57%	29%	0%	13%	0%	14%	50%
Internal project champions	88%	100%	71%	38%	0%	0%	0%	25%
Potential for Conflict among Participants	25%			0%	25%			88%
Number of issues dividing	88%	14%	0%	38%	0%	29%	43%	0%
Number of issues uniting	88%	71%	57%	25%	0%	0%	0%	13%
Level of agreement	13%	71%	43%	0%	13%	0%	0%	75%
Level of disagreement	13%	14%	0%	0%	0%	14%	43%	75%
Inherent value conflict	0%	14%	0%	0%	13%	14%	14%	50%
Inherent incentive conflict	75%	14%	0%	0%	0%	14%	29%	75%
Level of trust		57%	29%			29%	14%	
External Actors to Case								
External actors opposed	13%	14%	0%	0%	25%	14%	14%	38%
External actors in favor	50%	71%	29%	13%	0%	0%	0%	0%
Absence of external actors	25%	0%	0%	0%	13%	57%	29%	38%
Riskiness for external actors	13%	0%	0%	0%	25%	14%	43%	75%
External project champion(s)	63%	71%	14%	0%	0%	0%	0%	13%
Ext. demands, expectations	50%	14%	29%	0%	50%	29%	0%	25%
Media attention given project	75%	43%	29%	0%	13%	0%	0%	0%

*Positively and negatively influenced were determined by respectively +2 or +3 and −2 or −3 on the attribute influence scale. Quartile percentages can add to more than 100% due to the possibility of strongly conflicting ratings among the raters.

The potential for conflict is most clearly negative for the low quartile (as would be expected). Five of the seven attributes have half or more the cases negatively influenced. On the positive influence side, issues uniting, level of agreement, and level of trust are especially strong for the upper two quartiles. This would also be expected, and it is supportive of the motivating assumption to see it confirmed.

Given that the cases are related to wicked problems, it is perhaps surprising that the value and incentive conflict attributes were not more negatively influential even in the upper three performance quartiles. One explanation is of course that the processes helped manage these conflicts especially well. In-depth interview material is supportive of this explanation. However, other contributing factors are also likely. First, an action experiment would have likely not been proposed to TransForum if its initiating team was riddled with dissension on such fundamental points. Second, TransForum would not have approved an action experiment with a team in such a state of disarray. Both of these types of selection bias are likely to be present in the cases, and they suggest that the TransForum selection process operated reasonably well.

Again in the context of wicked problems, it is perhaps unexpected to see such little negative influence from external actors. For the low quartile, only the attribute on 'riskiness for external actors' had a negative influence on more than half the cases, specifically 75%. Four of the seven attributes were positively influential for the high quartile and three for the moderate high quartile. Again, one suspects the positive impact of the selection process. If the KENGi partners are present, there is less likelihood of negative influence from actors external to the action experiment. In-depth interviews revealed that in fact the upper quartile cases worked to create positive external champions and persuade other external actors to be supportive of the action experiment's aims and outcomes.

Across all three categories of membership, potential for conflict, and external actors, the motivating assumptions are supported by the findings related to assumption 4, multi-stakeholder engagement.

6.4.2.3. Influence of attributes related to motivating assumptions 2, 3, and 5

Motivating assumption 1 (sustainable development is a dynamic process) gave rise to the impact side of performance including six distinct attributes. Assumption 4 (multi-stakeholder engagement) has just been analyzed in the prior subsection. What remains is an analysis of assumptions 2 (sustainable development needs system innovation), 3 (system innovation is a non-linear learning process), and 5 (multi-stakeholder approaches imply trans-disciplinary knowledge creation). These assumptions did instruct several of the performance attributes, but now the focus shifts to the explanatory attributes related to them.

One might expect that the individual explanatory attributes could be assigned one by one to the assumptions they are most closely aligned with. It turned out that such an assignment process proved rather simple-minded. These three motivating assumptions have highly interrelated impacts on what has influence and on the strategies for creating the influence. As a result, they are dealt with collectively.

Table 6.6 presents the explanatory attributes related to these three motivating assumptions. The attributes are organized under two category headings: innovation management and impact of intermediary. These attributes form the core of the innovation and learning process central to the set of motivating assumptions.

Table 6.6. The influence of attributes related to motivating assumptions 2, 3, and 5 (innovation; learning; transdisciplinarity) on case performance.

	Percent of Cases Positively* Influenced by Attribute and by Performance Quartile				Percent of Cases Negatively* Influenced by Attribute and by Performance Quartile			
Attribute	**HQ**	**MHQ**	**MQ**	**LQ**	**HQ**	**MHQ**	**MQ**	**LQ**
Innovation Management								
How 3P impacts considered	100%	57%	29%	25%	0%	0%	29%	38%
Reconcile differing incentives	50%	43%	14%	13%	13%	14%	0%	50%
"Translation" work:								
to find common language	50%	57%	57%	25%	0%	0%	0%	13%
to meet incentive needs	50%	57%	29%	25%	0%	0%	0%	50%
How explicit knowledge used	63%	43%	57%	25%	0%	0%	0%	0%
How tacit knowledge used	50%	57%	43%	13%	0%	29%	29%	13%
Strategies for:								
Experimentation in action	100%	86%	86%	50%	0%	0%	0%	0%
Innovation	88%	29%	14%	0%	0%	0%	0%	25%
Action learning	75%	43%	14%	13%	0%	0%	0%	50%
Generating new knowledge	88%	57%	57%	25%	0%	0%	14%	25%
Trans-disciplinary collab.	100%	57%	57%	0%	13%	14%	14%	63%
Impact of Intermediary								
Function(s) assigned	75%	43%	57%	0%	13%	0%	29%	50%
How functions carried out	63%	43%	71%	25%	0%	0%	14%	50%
Function(s) changed	38%	71%	29%	0%	13%	0%	0%	38%
When in life of TransForum	50%	14%	43%	13%	0%	43%	43%	63%

*Positively and negatively influenced were determined by respectively +2 or +3 and −2 or −3 on the attribute influence scale. Quartile percentages can add to more than 100% due to the possibility of "strongly conflicting" ratings among the jurors.

For the high performance quartile, every one of the 11 innovation management attributes positively influenced 50% or more of the cases in the quartile – three attributes are at 100% and three more are at 75% or above. The strength of influence falls off in the moderate high quartile, but 10 of the 11 attributes still positively influenced 43% or more of the cases. Even in the moderate quartile, six of the 11 attributes have 43% or more of the cases positively influenced. For the low quartile, only three attributes were negative influential in 50% or more of the cases, specifically 50% for reconcile differing incentives, 50% for translation work done to meet incentive needs, and 63% for strategies related to trans-disciplinary collaboration. The influence levels of explanatory attributes in this category are as expected, and they are highly supportive of the effectiveness of the set of motivating assumptions.

The final set of attributes assessed by the jurors directly examined the influence of the action experiment's intermediary. The influence of intermediaries was broadly positive for the top three quartiles, but not overwhelmingly so. The intermediary experience was negatively influential for the low quartile. Again, recall that the low quartile has four of its eight cases selected prior to the motivating assumptions being in place and three more selected in the early phase of using the assumptions. The in-depth interviews reveal an appropriately reflective attitude regarding the intermediary role. It was a role among many in the life and progress of the action experiments. The intermediary could not and did not do the work of the action experiment. As a result, it could only have so much influence. The multi-stakeholder team which included the intermediary shouldered the work and got the credit for the work from the jurors.

Overall the explanatory attributes related to motivating assumptions 2 (the need of system innovation), 3 (non-linear learning) and 5 (trans-disciplinary collaboration) are appropriately influential on project performance. The effectiveness of the set of motivating assumption is supported further.

6.4.3. Key findings

The following listing summaries the key findings presented in the chapter:

- The performance of the cases ranged from a low of 2 to a high of 8.3 on a 2–10 point scale. The cases were grouped in four performance quartiles. Impact and process performance attributes moved together as they declined across the quartiles. These two major components of performance thus appear to be linked, consistent with the motivating assumptions. Triple P performance was well balanced within quartiles and declined systematically across quartiles. Impact performance did fall off for all quartiles as the individual attributes moved from planning to investment to harvest of actual returns.
- The differential performance of the seven pre-assumptions cases vs. the 23 early/late assumptions cases strongly supports the collective set of assumptions as substantially related to project performance.
- Of the 76 explanatory attributes, 45 influenced 50% or more the cases. The most influential attribute had influence in 28 of the 30 cases while the least influential had influence in three cases. As expected with wicked problems, the portfolio with action experiments has many and diverse influences.

- The 14 explanatory attributes related to general process management had substantial positive influence in the upper performance quartiles and negative influence in the low quartile. This finding suggests that standard project management processes and tactics cannot be ignored in innovation. The in-depth interviews however added an important caveat to this finding: the traditional processes must be made more flexible to account for the loop backs, restarts and responsiveness to reflection and learning that characterize innovation.
- The vast majority of the 27 internal context attributes were positively influential for the upper performance quartiles and negatively influential for the lower, while the 15 explanatory attributes related to external conditions were not highly influential. In balance, it still seems appropriately cautionary to conclude that the set of motivating assumptions is a contingent model. Deciding when and if to use it should be a matter of its fit to a specific case context, both internal and external.
- Initiating project and process conditions are strongly influential in the positive for upper quartile cases and in the negative for the low quartile cases. How, why and who initiated the action experiment are critical concerns that must be managed from the very start of an innovation in a wicked problem arena.
- Multi-stakeholder engagement (assumption 4) showed itself to be highly influential in regard to case performance. With the KENGi group engages in the action experiment, the relevant players are inside the experiment and not outside. As a result, the potentially negative influence of outsiders declines dramatically and the potential for conflict among participants appears to become manageable. The opposite effects are seen for the low performance quartile cases.
- The 11 explanatory attributes related to innovation management (assumptions 2, 3 and 5) are all strongly positive in their influence on the upper quartile performers and negative for the low quartile. The strategies related to experimentation, innovation, action learning, new knowledge generation and transdisciplinary collaboration appear to be effective.

Collectively this set of findings is highly supportive of the effectiveness of the set of motivating assumptions. When a sustainable development project is managed by a process implementing the set of motivating assumptions, that project tends to perform well in generating both desirable impact and process outcomes.

6.5. From motivating assumptions to a practical innovation model for innovation

Based on the research presented in this chapter so far, the overall conclusion is that:

The set of five motivating assumptions were largely supported as a viable model for managing action experiments dealing with the wicked problem of sustainable development.

In addition to applying the five assumptions in practice, we can also identify two core strategies that an action perspective for innovation on wicked problems also needs.

Core strategy one is about action experiment formation and guidance. Careful selection, process formation and guidance of the action experiment effectively addressed the managing of initiating as well as continuing conditions for the cases.

Core strategy two is related to process monitoring and reflection. The in-depth interviews strongly supported this strategy which involves dedicated process monitors and specially set-aside time for reflection as critical to the influence of the explanatory attributes. The practical innovation model we are suggesting consist thus of the five assumptions plus the two core strategies.

6.5.1. Core strategy I: Action experiment formation and guidance

In order to make the motivating assumptions real, the portfolio of action experiments was selected and guided by a strategy that evolved over time. Besides the first seven action-experiments, all action experiments were initiated by an interested party such as a researcher, an entrepreneur, a consultant or a provincial official. Within the selection of the action experiments in the TransForum program, extensive external consultation and reflection has been conducted to examine the relevance of the action experiment to sustainable development in the agricultural sector and to check if all relevant KENGi-partners were engaged in the multi-stakeholder team. Also, the action experiment's leadership had to be designated. Early on, this meant a project leader or manager. It appeared that as project leaders, entrepreneurs often brought a helpful energy level to the execution of the action experiments.

What is also unique in the guidance of the action experiments is that every action experiment also had a *process monitor*. The process monitor was responsible to monitor the innovation process of the action experiment and intervene to keep it on track. The second core strategy closely relates to this.

In-depth interviews make it clear that this strategy for project formation and guidance was core to explaining the strong influence patterns observed in the attribute assessments. When formation and guidance was fully present, it was strongly associated with high performance. This strategy was not a business-as-usual project process. Formation and guidance is done differently in order to enhance innovation in the wicked arena of sustainable development, and that difference appears to have been strongly influential in the positive when present and in the negative when absent.

6.5.2. Core strategy II: Process monitoring and reflection

The analysis of Table 6.6 provides ample evidence that the 'how-to's' and 'strategies for' had substantial influence. These findings however are of diminished use if every case had a unique set of how-to's and strategies for. The in-depth interviews gave many examples of the idiosyncratic nature of the various processes related to innovation, experimentation, action learning, new knowledge generation and trans-disciplinary collaboration.

Simultaneously achieving innovation, experimentation, action learning, new knowledge generation and trans-disciplinary collaboration seems obviously to be a deeply confounding task, especially within the context of a messy, contested wicked problem like sustainable development. To deal with that, in all the action experiments systematic attention was given to process monitoring and reflection as an element to the innovation management. This was done by a specific person in the team that focused on the innovation task exclusively, the process monitor. This systematic element is named here as the second core strategies – the strategy for process monitoring and reflection.

Monitors made bumps and obstacles in the innovation process visible and did not allow teams to ignore them or brush them aside. The bumps and obstacles were to be overcome with all partners engaged. The process monitor was critical to keeping innovation, learning and collaboration functioning within the multi-stakeholder teams. Their portfolio of intervention tactics ran to the reflective side – continually posing the 'why' question, mirroring the problem presenting itself, offering process advice as the opening with the field team arose, encouraging field leadership to act when needed, and empowering participants to deal with critical issues.

Many of the inventions will be unique to each case and its context. One of the favorite strategies was the 'reflection workshop' (see Hoes et al. 2010). Such a workshop was also used to periodically pull the field team together to reflect on the progress and learning to date and then set new directions or confirm existing ones based on the reflection. Reflection alone was never the goal. Reflection spurring new or renewed action was the intent – keeping the innovation trajectory alive or moving to a more productive one. One specific tool developed to support this reflection process is called the 'Dynamic Leaning Agenda' (see Section 6.6.4. for a description).

Altogether, the in-depth interviews all echoed a common theme: Without emphasis on monitoring and reflection, the multi-stakeholder teams would not have taken the time to do these processes – further evidence that the strategies changed normal operating procedures. The team often was unable to reflect, and only sometimes wanted to act without reflection or to avoid conflicts rather than take them on to resolve them. After the reflection workshops and related interventions, team members many times acknowledged the great value that the reflection process had been and that it would not have occurred without the discipline to do so.

The analysis strongly suggests that this second core strategy focused on process monitoring and reflection was an especially influential element in applying the motivating assumptions and what the multi-stakeholder teams would *not* have done without TransForum's urging.

6.6. Key learnings from working with the motivating assumptions and core strategies

To this point the chapter has focused on a special research project designed to examine the set of motivating assumptions based on a meta-analysis of the portfolio of action experiments. This has led to a practical innovation model consisting of the five motivating assumptions and two core strategies. In this section, some key findings for working with the practical innovation model are discussed. The findings come out of research and experiences in special learning projects that have been conducted in and complementary to the action experiments (see appendix I for the relation between the action experiments and the learning projects).

The learning projects support the action experiments in giving concepts and tools for reflecting and cooperation. This supports the findings during the experience of the action experiments and scientific research that supported the evolution of the motivating assumptions.

This section shares some of the key findings of these learning projects from working with the motivating assumptions and core strategies. The findings give insights in how to work with the motivating assumptions and core strategies. The insights are described as management challenges and interventions tool, based on the bottom-up experiences in the 30 action experiments. When combining these action-learning insights with the practical innovation model, we are able to give a first suggestion for a 'Traveler's Guide to Managing Sustainable Development', which we present in Section 6.7.

6.6.1. Dealing with the wickedness of sustainability in setting project vision and ambition

Sustainability is regarded as a wicked problem, for which no true solutions exist (see Chapter 1 and Peterson 2009). This makes sustainable development a dynamic process with multiple dimensions and values of the stakeholders involved. Working with different value sets in a multiple stakeholder setting requires special attention for these differences. For setting the vision and ambition of a project in a multi-stakeholder setting, progress in several action experiments only became possible when the different stakeholders acknowledged that the action experiment aiming to contribute to a more sustainable development of agriculture will never be un-debated. This is a crucial step to take, since it implies that no single actor or group of stakeholders, no scientific research but also no process of negotiation is able to create an ultimate sustainable solution. If this notion is disregarded all too often multi-stakeholder processes lead to endless deliberating and negotiating, ultimately leading to a situation where stakeholders dig in and start debating each other's value set instead of looking for possibilities that connect those different value sets.

Diepenmaat (2009) refers to this as the 'optimization paradox', the situation where every stakeholder strives for what they believe is 'good' or 'just'. Since there is no good or just in dealing with wicked problems, this can lead to combinations of different *individual* optimization processes creating a very suboptimal outcome and sometimes revolt on a societal scale. The other situation that is often created, is that no new practices are created at all, because stakeholders are unable to create an agreement since their value sets and thus their views on what sustainability means are perceived to be too far apart instead of just different views within the same palette.

Voltaire noted already that 'the perfect is the enemy of the good'. The utopian search for the ultimate sustainable solution is nowadays seen as one of the biggest obstacles to create meaningful progress in the sustainability domain. It has therefore been advocated that for these types of intractable (Schön and Rein 1994), wicked (Rittel and Webber 1973), unstructured (Hisschemöller and Hoppe 1996) or persistent (Loorbach 2007) problems, new types of strategies are needed. Central in these approached is the notion that wicked problems like sustainability are not solved, but can only be managed. In our experiences we learned that this is only possible if, in the multi stakeholder process and at the initiating stage of the action experiment, the different values of stakeholders are taken into account to see where they connect. Often, value sets of stakeholders do not differ that much, but it does require to open up for other views. In this way, one builds constructively on new value sets that can be used for working toward sustainable development. For this process of connected value development, tools are being developed. This exceeds the scope of this book at this moment.

6.6.2. Dealing with inherent complexity

Regeer (2010) describes that management strategies for dealing with the inherent complexity of sustainable development, such as those employed by mode-2 intermediaries, are characterized by non-linearity, mutual dependency and experimentation. The challenge is how to create a robust yet flexible process (Cash and Moser 2000) based on these characteristics. In their account of reflexive governance for sustainable development, Voβ et al. (2006) shed light on the so-called efficacy paradox. The efficacy paradox says that on the one hand to be able to act, complexity must be reduced. In the classical paradigm this was done by a linear process of problem formulation and goal setting and implementation through differentiation of tasks. However, this may easily lead to the neglect of long-term system effects. Yet, on the other hand, consideration of all possible effects, by fully embracing the complexities, interdependencies and value pluralism, reduces the capacity to act (Voβ 2006: 436). Based on research in the portfolio of action experiments, Regeer concludes that the efficacy paradox is best dealt with through a governance process that is characterized by a simultaneous opening-up (through active participation in projects) and closing-down (through reifying vision, approach and lessons from projects in books, articles, fact sheets, case descriptions, etc.).

6.6.3. Dealing with knowledge creation in multi-stakeholder groups

As described before, sustainability is a wicked problem and can thus not be solved, but only be managed. Because of the wickedness of the problem the key goals of this management approach are 'learning to learn' and 'co-creation of knowledge'. To do this the approach needs to stimulate reflection and needs to be transformative and responsive to emerging issues. Managing sustainability is thus an explorative process that needs to attend both impact and process. Regeer (2010) therefore describes the need for 'mode-2 strategies'.

Mode-2 strategies share a commitment to addressing complex sustainability problems by involving multiple actors (notably social and natural scientists, entrepreneurs, administrators and governors, activists, citizens) acknowledging the multi-level nature of the problems and articulating and connecting multiple perspectives. To actively involve different stakeholders some scholars state the need for a boundary organization mainly in terms of mediating *between* different actors, adjusting the actions of one to the needs of the other (for instance through funding requirements).

We have instead experienced in the action experiments that finding solutions to the problems of sustainable development requires close collaboration between different societal actors (including scientists) in order to generate knowledge that not only addresses societal problems but also contributes to their solution (action oriented, confirm mode-2). Regeer et al. (2009) describes a conceptual distinction between knowledge integration and knowledge co-creation. She argues, based on research in the action experiments, that a well guided action-learning process of knowledge co-creation can unlock a huge knowledge resource. This source of knowledge is very valuable in the innovation process and would have been neglected if knowledge had been considered as decontextualized and disembodied to be integrated and made fit to societal problems through careful but detached analysis (see also Peterson 2009 on new knowledge). In the action experiments, mode-2 approaches toward knowledge creation have been successfully applied. Mode-2 knowledge has five characteristics (Regeer 2010).

1. Mode-2 knowledge creation is carried out in the *context of application*, rather than transferred from science to society.
2. Mode-2 knowledge creation is *trans-disciplinary*, where the prefix 'trans' indicates that knowledge is not necessarily derived from pre-existing disciplines; it *trans*gresses boundaries. In order to solve problems, a range of theoretical perspectives and practical methodologies is mobilized within the context of application.
3. A third characteristic of mode-2 knowledge is its *heterogeneity*, which refers to the much greater variety of sites where knowledge is created (think-tanks, consultancies, NGOs) as well as to the heterogeneity of skills and expertise brought in by the multiple stakeholders involved.
4. A fourth characteristic of mode-2 knowledge is its *social accountability* and *reflexivity*. Knowledge creation is increasingly becoming a dialogical process between research actors and those who have traditionally been outside of the science system. Problems and the corresponding stakeholders with their interests and needs are no longer at the 'outside' of the knowledge creation process; they influence topic choice and research design. Hence, social accountability permeates the whole knowledge creation process.
5. Finally, mode-2 is characterized by novel forms of *quality control*. Quality is determined by a wider set of criteria (than those used in peer review), reflecting the broadening social composition of the review system. These characteristics are all in contrast with traditional mode-1 science, which is characterized as being homogeneous, autonomous and generated within a disciplinary, primarily academic context (Gibbons et al. 1994; Gibbons and Nowotny 2001; Nowotny et al. 2003).

Concerning to the third characteristic, the heterogeneity of knowledge creation, Hoes et al. (2008) indicates from research in several action experiments, that collaboration between different actors from various institutions does not necessary need a continuous high degree of interaction between these heterogeneous actors in order to have success. Instead she indicates action learning along multiple *homogeneous* plan-action-reflection loops, which interact at transaction spaces (points of interaction between heterogeneous stakeholders), to generate the needed heterogeneity. Hoes points out three key concepts for shaping this heterogeneous action learning in system innovation:

1. *homogeneous articulation*, which helps focus on the under-articulated needs of specific actor groups and facilitates subsequent heterogeneous interaction;
2. *contextualization*, which helps relate the project aims to the different institutional settings of participants;
3. *congruency*, which helps formulate the questions that are meaningful from different (institutional) perspectives (Hoes 2009).

Thus, when applying multi-stakeholder approaches in an action experiment, investing in action learning with stakeholders from similar communities appears to be fruitful for facilitating effective interaction between the stakeholders of different communities.

The role of science in mode-2 knowledge creation differs from knowledge creation in mode-1 science. Rather than providing knowledge and reflection in innovation processes, scientists interact with the other stakeholders in the process of knowledge creation. In the action experiments New Mixed Farm and SynErgie, scientists and entrepreneurs successfully interacted to mutually contribute to each other's questions and insights in the problem at hand (Hoes 2009). Analysis show that in successful action experiments the interaction between scientists and entrepreneurs is formed by mutual processes of planning, action and reflection (Hoes 2009).

According to Regeer (2010) the experience in the action experiments shows that this process of coproduction is not unambiguous. Providing a forum for coproduction of information (one of the requirements of a boundary organization) is not enough to face the hampering factors induced by the differences in culture, currencies, language and reputational system of each of the collaborating parties. What is needed is an intensive process of opening-up problem definitions, goal formulation and intervention strategies, involving frame reflection and competence development. Core strategy 2, monitoring and reflection, is in line with this requirement. Section 6.6.4 gives tools and methods that are developed for monitoring and reflection.

6.6.4. Dealing with monitoring and reflection

Applying the motivating assumptions in practice means simultaneously dealing with innovation, experimentation, action learning, multi-stakeholder collaboration and new knowledge creation. Core strategy 2 addresses the need for monitoring and reflection in managing these processes. In the execution of the action experiments we gained experience with the processes of monitoring and reflection.

Hoes (2009) describes the conditions that need to be in place to create a safe environment in which reflection is stimulated. She states that requirements for effective collaboration within action learning practices include conditions at both the action experiment level and the participant level. At the action experiment level, an important condition for collaboration is a safe social setting based on trust between the participants (Coughlan and Coghlan 2004). Also a balance between familiarity and excitement is important in developing a good learning environment (Wenger et al. 2002). Conditions at the personal level include being open towards new experiences and feedback, being willing and capable to change personal behavior (Raelin and Raelin 2006), and value multiple perspectives (Willis 2004). At the action experiment level, some tools for process monitoring and reflection have been developed. The following section discusses some of these tools.

6.6.4.1. Dynamic Learning Agenda

Dynamic Learning Agendas have been developed within the action experiments and learning projects to bring to light the difficult, tough issues that are normally 'swept under the rug'. These issues can become more visible when there is room for exploration and even acceptation of differences between people and positions. Reflecting on the Dynamic Learning Agenda in a safe environment aims at exactly that. Regeer (2010) describes that the learning agenda contains the issues (formulated as questions) a team struggles with at a specific moment in time. By constructing a sequence of learning agendas, the agenda becomes dynamic.

A Dynamic Learning Agenda is constructed by explicating the challenges as they are implicitly discussed in a meeting on the progress of an action experiment. The Dynamic Learning Agenda then forms input for the next meeting and can become a tool to structure the meeting and discussions on progress. Regeer (2010) states that like learning histories, the learning agenda should be judged by the quality of the conversation they provoke (Kleiner and Roth 1996). As such they are intended to contribute to the sensitization of participants to the issues that emerge as relevant.

By including perspectives of a wide range of participants (initiators, followers, and opponents of the sustainable development at stake) no one perspective on sustainable development is excluded in advance. Furthermore the Dynamic Learning Agenda is devised as a means towards learning and mutual sense making, and is thus inextricably linked to the intervention process.

6.6.4.2. Interactive learning and reflexive monitoring in action

Regeer (2010) states that as mode-2 programs are characterized by emergent design, learning process between different actors and adaptive management, managers of these kind of programs have new demands for evaluation and evaluators. Regeer (2010) meets this demand by describing a trans-disciplinary innovation strategy 'ILA' (interactive learning in action) that has been developed during the 80s and 90s of the last century by Bunders (1990) and Bunders and Broerse (1991). The approach is structured around a number of principles stating the importance of involving end-users from the start, striving for knowledge co-creation, organizing a network of support and facilitation by an interdisciplinary team. The intervention strategy is cyclical whereby tailor made interventions follow observation of and reflection on previous interventions.

Broerse (1998) formulated the importance of recognizing that conditions for a mode-2 strategy are never in place and that, as a consequence, the strategy itself should focus on creating or dealing with these conditions. Thus, the ILA approach does not only act at project level, but also on the institutional setting and the individual competences of people.

Based for a considerable part on the experiences in the action experiments, the ILA-approach has been developed into an approach of 'reflexive monitoring in action' (RMA). RMA is an action-oriented monitoring approach. The monitoring activity itself is with RMA not a separate activity, but an integral part of the process. RMA stimulates project participants to constantly reflect on the relationships between the most important issues: the project ambition, the current realities and practices, their institutional embedding and the developments in the system that offer opportunities to realize the system innovation ambitions (Van Mierlo et al. 2010b).

6.6.5. Dealing with the tension between a project and its institutional context

Monitoring and reflection is an important element of innovation management. The former section addressed point of attention and tools for dealing with monitoring and reflection between the stakeholders within a project. Another important point of attention is monitoring developments between a project and its external environment. Implementation of innovation is a crucial aspect for success, as is stressed more often in this book, and so are dealing with the barriers to come to implementation. Research by Klerkx et al. (2010) conclude that because innovation networks can only partially influence their institutional environment, and because unintended consequences of actions and random events influence the course of the innovation process, innovation network actors need to continuously reinterpret the contexts in which they move. This constant reflection by the innovating actors on their position vis-à-vis their environment needs to be supported by dedicated facilitators and monitoring and evaluation methods aimed at system learning. This implies that agricultural innovation policies should, instead of aiming to fully plan and control innovation, foster the emergence of such flexible support instruments that enable adaptive innovation management.

An important outcome of the study by Klerkx et al. is that innovation networks may support their effective reformism efforts by using different types of boundary spanning actors to defend and advocate their interests, broker new contacts, and mediate in case of conflict. Furthermore, using tangible visions and artifacts helps them to create understanding about and support for their ideas (see also Chapter 4). However, when these visions become too rigid, they may lead to the innovation process becoming locked-in (Klerkx et al. 2010).

Both internal project factors (e.g., new actor involvement or actor exits) and external project factors (e.g., economic developments, policy discontinuities) can induce both positive interaction cycles (decreasing uncertainties) and negative interaction cycles (increasing uncertainties) that influence the agency of innovators (Klerkx et al. 2010). The occurrence of such positive and negative interaction cycles show that adaptive management implies that innovation networks need to keep a keen eye both on the dynamics within their innovation network, and developments at different places and innovation system levels (c.f. Blay-Palmer 2005) in their direct institutional environment and beyond. This implies that innovating actors need to continuously assess how they can realize their innovation goals within the ever changing institutional environment, looking for certain 'windows of opportunity' and sometimes also trying to actively create such windows of opportunity (Klerkx et al. 2010).

In terms of making innovation policies adaptive, as opposed to a rigid planning and control orientation, the importance of having innovation system facilitation becomes apparent. According to Klerkx and Leeuwis (2009) a promising approach seems the employment of specialized innovation brokers (such as TransForum). Such innovation brokers assist in vision formulation and reformulation, continuous network formation and adaptation, and facilitation of multi-stakeholder interaction by means of network coordination and mediation (Klerkx and Leeuwis 2009; Kristjanson et al. 2009), by applying monitoring and evaluation methods which are aimed at learning (Horton and Mackay 2003). Given the interaction between different system levels in effective reformism efforts, such methods need to support continuous reflection on the micro position of the innovation networks and their goals versus macro-level systemic possibilities and constraints. Methods such as reflexive process monitoring incorporating innovation system failure analysis (Van Mierlo et al. 2010a), the interactive learning and action monitoring approach (Regeer et al. 2009) and participatory impact pathway analysis (Douthwaite et al. 2003), all designed to pro-actively create awareness amongst innovation network actors as regards their position vis-à-vis their institutional environment, seem promising in this regard (Klerkx et al. 2010).

6.7. Applying the TransForum model

6.7.1. A traveler's guide to managing sustainable development

A traveler on the road to sustainable development stops to ask directions. What does anyone connected to the TransForum experience say in response? For now, it would seem that the following guideposts, based on the five motivating assumptions, the two core strategies and the key learnings from working with the practical innovation model, constitute a provisional guide to managing sustainable development:

- Manage initiating conditions through an effective strategy for project formation and guidance. Have the KENGi partners engaged from the start and keep them engaged by organizing a relatively safe internal environment and constantly challenging and inspiring them via active networking.
- Practice sound general project management with the caveat that such management must be applied more flexibly and adaptive with innovation projects,

most especially in the areas of control processes and schedules. Allow the needed loop backs, restarts, and adjustments in a dynamic process.
- Align and fit the specifics of the management strategy to the external and internal contexts of the situation. Of the 76 explanatory attributes, 45 were strongly influential in half the cases or more. The arena of wicked problems is complex and the critical success factors are many and likely tied to the specific situation.
- Practice multi-stakeholder engagement (assumption 4) broadly and continually in project work. Do not sweep difference under the rug. Challenge and engage. Acknowledge the different needs of the various stakeholders and their institutional context. Collaboration and innovate together. Co-create means just that.
- Implement system innovation (assumptions 2), non-linear learning (assumption 3), and trans-disciplinary collaboration (assumption 5) through an effective strategy for process monitoring and reflection. Learning and innovation seem to be unnatural acts for many. Make them natural through the processes used.
- Focus on a multi-dimensional, dynamic set of performance attributes that measure both impacts on the target system and progress in the process of stakeholder engagement. Perhaps this one should have been listed first. But in reality, these guideposts are not a list. They are a continuously evolving spiral.

6.7.2. Implications with regard to the current innovation system

The findings of our research are only briefly presented here. These findings both support and extend the findings from the meta-analysis reported in the earlier sections of the chapter. Managing the processes related to the five motivating assumptions again shows itself to be a complex and dynamic system of strategies and interventions. Great care is needed if the process management is to result in both system impacts and healthy process outcomes.

Collectively these findings raise several concerns about the existing knowledge infrastructure and agro innovation. The more traditional system appears to lead in directions that are opposite to the ones suggested by the set of motivating assumption (or at least do not stimulate the wider application of the assumptions). Some of these more vexing counter directions include:

1. Still trying to define and measure sustainability in any conventional way
2. Over focusing on technological inventions, and neglecting the need for simultaneous and parallel software and orgware innovations
3. Prescribing (for instance in governmental rules for subsidies to projects) a traditional project management approach, and making projects accountable to the original planning, problem formulation and described deliverables, thus minimizing the room for adaptive management and emergent design.

4. Minimizing the heterogeneous co-creative space(s) and the existence of new intermediaries where KENGi-partners can come together and work together to try-out new things, without being completely accountable to the rules and conditions of the current regime.
5. Low valuation of applied and co-created knowledge in the scientific domain, which is dominated by scientific peer reviews and takes little account in the societal and business value of the new knowledge.

6.7.3. Broader application of the model

However positive the conclusions from the findings are concerning the motivating assumptions, the conclusions are based on only 30 cases set in a specific context. The proper next question is whether the findings hold when the context is shifted. In other words, how generally can the set of motivating assumptions be applied?

Among the findings there is reasonable evidence the motivating assumptions may be widely applicable to wicked projects in sustainable development. First, the set of motivating assumptions appears to be more successful with more wicked projects than less wicked. This further suggests that in applying the set of motivating assumptions to new settings the application should be to a wicked problem. Tamer problems do not appear to need the added process architecture of the set of motivating assumptions.

Second, it appears that the set of motivating assumptions is about equally suited to action experiments in mainstream agriculture and in niche agriculture. Applying the set of motivating assumptions on both sides of the scale divide is a useful insight into the general applicability of the model. Sustainable agri-food systems likely need effective innovation for both scales of agriculture as opposed to assuming one must supplant the other.

Third, a variety of innovation types were seen in the 30 action experiments. Although this finding needs much additional research, the first results suggest that the set of motivating assumptions worked reasonably well in all types. Again, this finding supports more general application of the model.

Finally, the action experiments were nearly all located and executed in the Netherlands. It remains a bit unclear how confining this fact is in regard to broad applicability. On the one hand, the problems found in the action experiments have many parallels in other mature agri-food systems. Other European nations, the United States, and the more developed economies of the Pacific Rim all have similar sets of problems and issues related to sustainable development. The experience appears quite general as seen from this perspective. Even in regard to developing agri-food systems, the focus of these systems on exportable commodities and products to developed markets and the emergence of global grocery chains standardizing procurement practices worldwide suggest that many of the issues encountered by TransForum also have their parallels in less developed settings.

Several clear limits to application are also rather obvious from the analysis. First, as qualitative grounded theory research, the findings have not arisen from a large expanse of cases and from statistical inference. Therefore, any additional application needs to be done with caution.

Second, none of the action experiments were done in an extreme external context with either exceptionally positive external conditions or exceptionally negative ones. There was no action experiment for example, in which the KENGi partners had previously taken firm negative public stands specifically against each other in relation to the project being proposed. It is not clear that the set of motivating assumptions would be useful in a polarized setting. In fact, none of the cases suggest that a highly effective innovation process can overcome a highly charged negative context. Quite the opposite, in the vast majority of action experiments, the context was reinforced by process rather than being inherently at odds.

Third, all of the action experiments were done in the agri-food system. The special issues of food, nutrition, safety and living biology may make the agri-food system a unique setting for wicked problems. As a result, applications of the set of motivating assumptions in non-agri-food settings may be a significant concern. Application of the assumptions and core strategies to such a non-agri-food setting would seem to be an ideal next experience in applying the model.

Fourth, TransForum is itself a rather unique intermediary. TransForum had substantial funds which the applying action experiments needed to achieve their goals. The funding gave TransForum a special status in creating the disciplines and processes surrounding the assumptions and core strategies. If the funding had not been there or if an organization like TransForum had not existed, would the action experiments have found an internally self-reinforcing mechanism for multi-stakeholder formation and process reflection? The answer from these 30 cases is largely no. TransForum's existence and funding was critical. Perhaps then the biggest question of general application relates to determining the circumstances under which such a unique intermediary (boundary organization) would exist.

Finally, across the action experiments, the weakest attributes of performance were the actual investment in and harvesting of value. Does real triple P value creation result from the TransForum model? The trajectory of the higher performance projects says yes, but the proof of value creation is left undone. Six years was not enough time to evaluate this. It probably shouldn't surprise anyone that when sustainability is concerned 6 years just is not enough time to judge.

6.8. References

Blay-Palmer, A. (2005). Growing innovation policy: the case of organic agriculture in Ontario, Canada. *Environment and Planning C: Government and Policy* 23, 557–581.

Broerse, J.E.W. (1998). *Towards a new development strategy. How to include small-scale farmers in the biotechnological innovation process.* Delft: Eburon.

Bunders, J.F.G. (Ed.). (1990). *Biotechnology for small-scale farmers in developing countries. Analysis and Assessment procedures.* Amsterdam: VU University Press.

Bunders, J.F.G. and Broerse, J.E.W. (1991). *Appropriate Biotechnology in Small-scale Agrciulture: how to orient research and development.* Wallingford: CAB International.

Cash D.W. and Moser, S.C. (2000). Linking global and local scales: designing dynamic assessment and management processes. *Global Environmental Change* 10, 109–120.

Coughlan, P. and D. Coghlan (2004). Action learning: towards a framework in interorganisational settings. *Action Learning: Research and Practice* 1(1): 43–61.

Diepenmaat, Henk (2009). *Een filosofie van de maatschappelijke praktijk.* Almere: Parthenon.

Douthwaite, B., Kuby, T., van de Fliert, E. and Schulz, S. (2003). Impact pathway evaluation: an approach for achieving and attributing impact in complex systems. *Agricultural Systems* 78, 243–265.

Gibbons, M. and Nowotny, H. (2001). The potential of transdisciplinarity. In J.T. Klein, W. Grossenbacher-Mansuy, R. Haberli, A. Bill, W. Scholtz and M. Welti (Eds.), *Transdisciplinarity: Joint Problem Solving among Science, Technology and Society. An effective way for managing complexity* (pp. 67–80). Basel: Birkhauser Verlag.

Gibbons, M., Limoges, C., Nowotny, H., Schwartzman, S., Scott, P. and Trow, M. (1994). *The new production of knowledge: The dynamics of science and research in contemporary societies.* London: Sage publications.

Hisschemöller, M. and Hoppe, R. (1996). Coping with intractable controversies: the case for problem structuring in policy design and analysis. *Knowledge and Policy* 8(4), 40–60.

Hoes, A.-C., Regeer, B.J. and Bunders, J.F.G. (2008). Transformers in knowledge production: building science-practice collaborations. *Action Learning: Research & Practice* 5(3), 207–220.

Hoes, A.-C., Regeer, B.J., Bunders, J.F.G. (2010). Facilitating learning in innovative projects: Reflections on our experiences with ILA-monitoring. *Paper presented at the International Conference on Organizational Learning, Knowledge and Capabilities OLKC 2010* (Boston, 3–6 June 2010).

Horton, D. and Mackay, R. (2003). Using evaluation to enhance institutional learning and change: recent experiences with agricultural research and development. *Agricultural Systems* 78, 127–142.

Kleiner, A. and Roth, G. (1996). *Field manual for the learning* ^{His}*torian.* MIT, Center for Organizational Learning. Cambridge, MA: Massachusetts Institute of Technology.

Klerkx, L. and Leeuwis, C. (2009). The emergence and embedding of innovation brokers at different innovation system levels: insights from the Dutch Agricultural sector. *Technological Forecasting and Social Change* 76, 849–860.

Klerkx, L., Aarts, N. and Leeuwis, C. (2010). Adaptive management in agricultural innovation systems: The interaction between innovation networks and their environment. *Agricultural Systems*, 103(6), 390–400.

Kristjanson, P., Reid, R.S., Dickson, N., Clark, W.C., Romney, D., Puskur, R., MacMillan, S. and Grace, D. (2009). *Linking International Agricultural Research Knowledge with Action for Sustainable Development* 9, 5047–5052.

Loorbach, D. (2007). *Transition management. New mode of governance for sustainable development.* Utrecht: International Books.

Nowothy, H., Scott, P. and Gibbons, M. (2003). " 'Mode 2' revisited: the new production of knowledge". *Minerva* 41, 179–194.

Peterson, H.C. (2009). Transformational supply chains and the 'wicked problem' of sustainability: aligning knowledge, innovation, entrepreneurship, and leadership. *Journal on Chain and Network Science* 9(2), 71–82.

Raelin, J.A. and Raelin, J.D. (2006). Developmental action learning: toward collaborative change. *Action Learning: Research and Practice* 3(1), 45–67.

Regeer, B.J. (2010). *Making the invisible visible. Analysing the development of strategies and changes in knowledge production to deal with persistent problems in sustainable development.* Oisterwijk: Boxpress.

Regeer, B.J., Hoes, A.-C., Amstel,van Saane M. van, Caron-Flinterman, F. and Bunders, J.F.G. (2009). Six guiding principles for evaluating mode-2 strategies for sustainable development. *American Journal of Evaluation* 30(4), 515–537.

Rittel, H.W.J. and Webber, M.M. (1973). Dilemmas in a general theory of planning. *Policy Sciences* 4(2), 155–169.

Schön, D. and Rein, M. (1994). *Frame reflection: toward the resolution of intractable policy controversies.* New York: BasicBooks.

Strauss, A.L. and Corbin, J. (1990). *Basics of qualitative research: Grounded theory procedures and techniques*. Newbury Park, CA: Sage publications.

Van Mierlo, B., Leeuwis, C., Smits, R. and Woolthuis, R.K. (2010a). Learning towards system innovation: evaluating a systemic instrument. *Technological Forecasting & Social Change* 77(2), 318–334.

Van Mierlo, B., Regeer, B.J., Amstel, M. van, Arkensteijn, M., Beekman, V., Bunders, J.F.G., De Cock Buning, T., Elzen, B., Hoes, A.-C. and Leeuwis, C. (2010b). *Refelxieve monitoring in actie. Handvatten voor de mointoring van systeeminnovatieprojecten.* Oisterwijk: Boxpress.

Voβ, J.-P., Bauknecht, D. and Kemp, R. (2006). *Reflexive governance for sustainable development.* Cheltenham, UK: Edward Elgar.

Wenger, E., McDermott, R., et al. (2002). *Cultivating communities of practice*. Boston: Harvard Business School Press.

Willis, V.J. (2004). Inspecting cases against Revans' 'gold standard' of action learning. *Action Learning: Research and Practice* 1(1), 11–27.

Yin, R.K. (2009). *Case study research: Design and methods.* 4th Edition. Applied Social Research Methods Series, Vol. 5. Thousand Oaks, CA: Sage publications.

Chapter 7. Transforming agriculture: A new approach to stimulate sustainable development

Karin Andeweg[1] and Henk C. van Latesteijn[2]

[1]Andeweg@TransForum.nl; TransForum, P.O.Box 80, 2700 AB, Zoetermeer, The Netherlands

[2]VanLatesteijn@TransForum.nl; TransForum, P.O.Box 80, 2700 AB, Zoetermeer, The Netherlands

Abstract In this final chapter we propose the TransForum model. The analyses based on the five motivating assumptions lead to the conclusion that they indeed have a positive impact on innovation initiatives that aim to improve the 3P performance, or in other words, form a next step on the road toward sustainable development. For that reason we propose them as principles that together constitute the framework of the TransForum model. Multiple observations lead to the formulation of a sixth principle that underlines the need for business models that incorporate the 3P aspects in a transparent and accountable way. Observations from the different disciplinary analyses add context and indications for operational tools and interventions for the two core strategies that accompany the six principles. We regard the TransForum model a first attempt to describe in a practical sense the possibility to manage the wicked problems of sustainable development in agriculture.

7.1. Introduction

In this book, the cyclical and fuzzy process of agricultural innovation that is needed for a transition toward a more sustainable development is analyzed. The sustainability challenge that the agricultural sector faces puts a real strain on the innovation process, because agriculture must find new ways of connecting with its environment. For that to happen, innovation is crucial. Not the type of innovation that we witnessed the last decades, but innovation that really contributes to a more sustainable development of the sector itself and society at large. We need an agricultural sector that is economically viable, environmentally efficient, and socially acceptable. The research that is described in this book took place under the umbrella of the TransForum program: an innovation program in The Netherlands aimed at opening up new perspectives for the agricultural sector and improving the meaningful application of knowledge, techniques, skills and organizational capacities that are available. The program focused on investigating new possibilities of performing agriculture within the new context of sustainable

H.C. van Latesteijn and K. Andeweg (eds.), *The TransForum Model: Transforming Agro Innovation Toward Sustainable Development*, DOI 10.1007/978-90-481-9781-1_7,

development and the approach taken can be best described as 'learning-while-doing'. The main research question that we address in this book is therefore formulated as follows: What is needed to transform a linear innovation approach based on the assumption that the existing knowledge infrastructure generates the necessary knowledge, to a new agro innovation system in which meaningful knowledge is generated by multiple sources and that provides a real contribution to a more sustainable development?

Based on earlier experiments and on research in the field of (agricultural) innovation we hypothesized that the answer to this question was to be found in organizing a process of engagement between different stakeholders. Searching answers related to achieving a more sustainable development leads to the observation that we have to deal with so-called wicked problems. This is discussed in more detail in Chapter 1. In short: wicked problems lack a clear definite formulation of the problem and consequently do not give rise to a clear solution. Therefore, innovation aimed at sustainable development of agriculture will accordingly not have one clear recipe for action. Every challenge is different and therefore needs a different approach. Academic and popular literature show broad analyses of the problem at hand. Norton (2005), Van Bueren et al. (2003) and others propose conceptual strategies for innovation in the field of sustainable development. However, most of these analyses focus on the general structure of the challenge and therefore lack a clear perspective for action.

For this reason TransForum has chosen an experimental approach so the result could be a set of guidelines or action perspectives that can be applied to concrete innovation initiatives. This needed an expansion of the current analytical scope to take it a few steps further and translate generic insights into actionable items. For example: one of the three principles that Norton (2005) proposes tells us that experimentation is needed to deal with the wickedness of sustainable development. But the question still remains how this experimentation can best be executed. Based on the first experiences in the field of agricultural development and notions from literature we formulated five motivating assumptions that helped us in the actual execution of a number of action experiments. Rather than further analyzing the problem and theorizing on possible adequate solutions, we applied these assumptions in practice.

The five motivating assumptions are not the definitive answer to the challenge. However, they reveal the contours of a framework to achieve progress in the field of sustainable development. By using these assumptions to execute a diverse set of action experiments that explicitly aim at a more sustainable development we were able to test the usefulness and validity of the assumptions. Complementary scientific research helped to better understand the problem and give insight on the 'how to' challenge.

In this final chapter we evaluate and summarize the results and mould them into a practical and applicable model to stimulate innovation in the field of sustainable development.

7.2. The TransForum model

In Chapter 6 Peterson and Mager presented a first outline of the TransForum model. The model comprises five validated motivating assumptions and two core strategies. The validation is based on a qualitative analysis of 30 action experiments that were executed using the assumptions. In this final chapter we expand the model by adding a sixth principle based on the observation in the various experiments and research efforts.

7.2.1. From assumptions to principles

The qualitative analysis in Chapter 6 reveals that the five assumptions were without any exception relevant in improving the performance of the action experiments. By applying them to very different types of innovations in action experiments it became clear that the assumptions indeed addressed the right practical implementation of the complex adaptive strategies that were suggested in the literature. Additional material from a number of learning projects also supported the notion that indeed the assumptions did lead to significant results. Furthermore, the scientific analyses that investigated the potential of influencing inventions (Chapter 2), the possibilities and blockages to organize innovations (Chapter 3), the role of images of sustainability in developing new businesses (Chapter 4) and the possibilities of using demand for sustainable consumption as a driver (Chapter 5) all point in the same direction.

We therefore recommend promoting the assumptions to principles to be applied in innovation trajectories aiming at a more sustainable development in agriculture. A few considerations and warnings should also be mentioned. A first consideration is that applying the principles and consequently the strategies, tools and project management processes that they imply, will only be useful if we are dealing with an innovation trajectory in the field of sustainability that reveals traits of a wicked problem. The principles are partly devised to address these traits and partly to define a contingent approach that overcomes the managerial difficulties of wicked problems. A second consideration is that the principles give you guidelines as how to go about, not a strict set of rules or a clear recipe for every imaginable situation. We feel comfortable enough to call them principles, since we have experienced that if you apply them, they are useful. However, bear in mind that all this is contextual, and therefore reflection and adaptation will probably be necessary with every new application. And finally a warning: our findings are based on a limited set of experiments and an evenly limited set of in depth analyses. Therefore the results should be seen as recommendations for further exploration. With this in mind, the proposed model is not a final result, but a starting point for a new line of research and development.

The principles are:

Principle 1: Sustainable development is a dynamic process

In all action experiments we had to deal with the dynamics of different and varying values. We applied and developed various tools such as *reflexive process monitoring*, *interactive learning and action monitoring*, and *participatory impact pathway analysis* to deal with these dynamics. Together these tools represent a specific kind of process management that accommodates an important aspect of wicked problems, i.e. the different values brought to the table by the various stakeholders. The tools and interventions were all aimed at connecting these values in a new mode of doing agricultural production, not by sharing them or re-shape them into one overarching value. A more extensive description of these process tools and their applications will be published separately. So, this principle addresses the dynamic process of sustainable development and helps to manage it in a practical way.

Principle 2: Sustainable development needs system innovation

Analyzing inventions for sustainable development – as described in Chapter 2 – led to the conclusion that applying hardware inventions will not bring about the needed transformational change. Also institutional challenges will have to be faced. This calls for innovations in the actual agricultural mode of production, but also for innovations in how we deal with these new modes in terms of personal and institutional competencies and organizing capacity. If this happens, the system as a whole will change. So, this principle states that every step into the direction of a more sustainable development in agriculture will inevitably lead to changes of the system in which agriculture is taking place.

Principle 3: System innovation is a non-linear learning process

Using system innovation to deal with sustainable developments is one way of handling or managing the underlying wicked problems. Because of the wickedness of the problem, the key goals of this management approach are 'learning to learn' and 'co-creation of knowledge'. The management approach needs to stimulate reflection and needs to be responsive to emerging issues. Managing sustainability is therefore a non-linear and explorative process that is aimed at both organizing the wanted impact and see to it that the process indeed leads to reflection and learning. So, this principle leads to the conclusion that dealing with sustainable development in agriculture calls for a well balanced attention between a project that is focused on attaining the desired goals, and a process that sees to it that sufficient reflection and adaptation is available to effectively respond to emerging issues.

Principle 4: System innovation requires a multi-stakeholder approach

Transformational change and a non-linear learning process imply input from different actors. The creation of new knowledge that underpins innovation toward a more sustainable development of agriculture needs to be carried out in the context of its application. This type of new knowledge production is characterized by a great variety of sites where the knowledge is created and a great variety of skills and expertise brought in by the multiple stakeholders involved. Knowledge creation for this type of innovation is increasingly becoming a process of engagement between researchers and other actors who have traditionally been outside of the knowledge production system. So, this principle leads to the inclusion of relevant stakeholders in the design and execution of the innovation process to enable the creation of new knowledge to occur.

Principle 5: Multi-stakeholder approaches imply trans-disciplinary knowledge creation

In a multi-stakeholder environment, boundaries between disciplines and other sources of information and knowledge will dissolve. In order to formulate an answer that connects to multiple visions, beliefs and values, a range of theoretical perspectives and practical methodologies has to be mobilized within the context of application. This is the essence of trans-disciplinary work where knowledge is created or constructed by transgressing boundaries between disciplines and fields of expertise. The appraisal of the outcome of such a process is not restricted to the standards and norms of academic, professional or societal peer communities. The final assessment will be made using multiple criteria applied by different value driven stakeholders that together will decide on the positive contribution of the new knowledge and its application to a more sustainable development of agriculture. So, this principle draws attention to the non-standard development and appraisal of new ideas, innovations and modes of agricultural production that surpasses the regular professional divisions in society.

7.2.2. The sixth principle

While applying the five principles in action experiments it became clear that a sixth principle is lacking. This sixth principle has to deal with the fact that innovation, at the end of the day, is about setting up a new business that can deliver on all three aspects of sustainable development. So, the new business should not only address societal and market wishes related to physical and societal environmental aspects, it should also lead to a value proposition that delivers enough profit to compete, or rather outperform current modes of doing agriculture. The comparison of the performance of the 30 action experiments given in Chapter 6 illustrates that only in a few cases actual implementation of an innovation was realized. Most of the action experiments within the program were aimed at developing a plan that was to be implemented outside the scope of the program. However, in most of the experiments the phase of developing plans marked the end of the action experiment. We came to realize that these phases, planning and investing and implementing should not been seen separate. Actual investment and implementation is needed to really assess whether an innovation will lead to profit and prospects. It is the notion of profit that ultimately will lead to actual implementation and action. And it is the same notion that is necessary to develop a self-reinforcing system of continuous innovation.

There are multiple lines of reasoning that led us to formulate the sixth principle. In Chapters 2 and 5 the importance is underlined of bringing new products that are produced with the innovative production system to the market. Unfortunately a considerable proportion of new ideas and new products eventually fail because there is no reward to be gained for it in the marketplace. In Chapter 2 Jacobsen, Beers and Fischer argue that next to innovations in hard-, soft- and orgware elements, entrepreneurship is a crucial aspect that should be taken into account when attempting to realize an innovation. A sound business model of the new product or service is a crucial part in the innovation process. In Chapter 5 Van Trijp and Fischer support this notion from a marketing perspective. They conclude from their research that the actual consumer demand for sustainably produced products is currently too weak. Therefore, the market does not self-regulate sustainable development at this moment. Market creation for sustainable development, or 'enlarging the pie' as the authors call it, is a collective effort that the different stakeholders need to incorporate in the innovation process. A sound business model is needed to 'push' the development, before a pull from society will be able to take over the driving force for the innovation. So, next to the five principles, we have formulated a sixth one to close the loop of innovation toward a more sustainable development.

Principle 6: New business models based on new knowledge lead to better 3P performance of agriculture

The essential elements in these new business models are the inclusion of people and planet aspects next to the regular profit aspects in a transparent and accountable way. It must be clear from the business model what the contribution to these three domains will be. And by presenting them transparent and accountable, the implementation of the innovation can be truly assessed on its performance. Although this all seems rather straightforward, in our action experiments, and for that matter in real life, we have experienced that actually doing it and take the original idea of the innovation to its consequences is a daunting task.

Together the six principles generate a framework that presents guidelines for the practical implementation of new modes of agricultural production that will have a positive contribution toward a more sustainable development. The framework is presented in Figure 7.1.

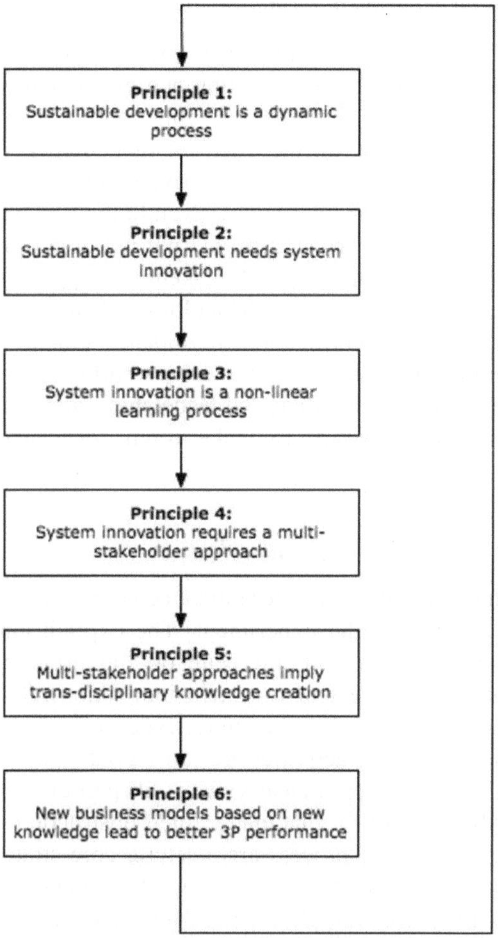

Fig. 7.1. The six principles that together constitute the TransForum model framework.

7.2.3. Accompanying strategies

Principles alone are not enough to shift from analysis into action. In Chapter 6 Peterson and Mager also indicated that in the execution of the action experiments two distinct strategies were visible. And furthermore, they conclude that without these strategies the application of the principles would probably not have led to positive results. The results from a number of additional studies aimed at understanding the learning process within the experiments and the effectiveness of different interventions supported the relevance of the strategies.

Core strategy 1: Action experiment formation and guidance

The formation of the experiments proves to be very important for its final performance. A deliberate process of selection and evaluation was needed to see to it that especially principles 1–4 were thoroughly applied. This hold particularly for the selection or attraction of the relevant stakeholders from knowledge institutes, entrepreneurs, societal groups and governments. If any of these were missing from the line-up, extra attention was given in the first phase of the experiment to mend this omission. When it appeared not possible to get all the needed stakeholders at the table to ensure the multi-stakeholder approach, the project was either terminated deliberately or ran into difficulties and stopped by itself. Another equally important aspect of the formation and guidance strategy is the designation of different roles and responsibilities to the various participants in the innovation effort. Not only should it be clear where the leadership, monitoring and execution is positioned, it is also necessary to make room for changes in roles and responsibilities over time.

Core strategy 2: Process monitoring and reflection

In next phases of the experiments further guidance comprised the active involvement of a process monitor in the experiment. The explicit aim of the monitor function is to observe, reflect and intervene on the process development in the experiment. The development of a system innovation requires the different stakeholder to get actively involved in learning, experimentation, debate, business development. And most of these activities are carried out simultaneously. The role of the process monitor is to keep a clear view of all participants to their original goal and their contribution – or for that matter, their obstruction – of the collective innovation process. In terms of process architecture, core strategy 2 is explicitly aimed at safeguarding principles 5 and 6. So, monitoring and evaluation is not an add-on activity, but fully accountable for realizing new knowledge creation in a trans-disciplinary setting and using this knowledge to develop a new business model that indeed delivers the anticipated improvement on the 3P performance.

Together, the six principles and the two strategies comprise what we would like to entitle the TransForum model. This model describes an executable and replicable way of setting up an innovation trajectory in a situation where the underlying wicked problem (in this case sustainable development) does not allow for more traditional ways of executing a more standard project approach. The extra effort to account for the wicked nature of the problem, expressed in different value orientations and consequently different ideas about potential solutions, consists of a mix of concise project management and appropriate process monitoring and intervention.

7.3. Contributions from scientific disciplines

The analyses of inventions for sustainable development, of organizing innovations, of images of sustainable development and of stimulating sustainable consumption have all given additional insights in the application of the six principles. Next to that, these analyses give additional information on the use of tools and approaches within the two core strategies. In the various chapters contributions from very different scientific disciplines are given that together may help the transformation of the current agro innovation system. Two broad approaches can be discerned: looking through the lens of natural sciences and the lens of social sciences. A summary of the observations and the conclusions that were drawn is given in the nest paragraphs.

7.3.1. The Natural Sciences lens

In Chapter 2 the role of inventions and the possibility to stimulate these inventions into a given direction, is described from a technical sciences perspective. Jacobsen, Beers and Fischer argue that inventions were an important aspect of agro innovation. And they will remain important in the future. This concerns both new inventions and new combinations of existing inventions. In a number of examples this was made clear.

The new double challenge to innovation – both improving the operational efficiency and showing a better track record with respect to planet and people values – does lead to the need for inventions in different domains. First of all the classical invention that lead to new piece of hardware is still necessary. The hardware can relate to new modes of production, but also services and products. Because these new hardware entities will also need to transform the way in which agriculture is being executed, innovations in software are also needed. This means that competencies and skills of people dealing with the new hardware also will be renewed. And moreover, the new hardware that is operated in a new and different way will also need to become part of a more encompassing agricultural production system. For that to happen, inventions in 'orgware' are needed as well, meaning that existing organizations, institutions and regulations will need to be redesigned simultaneously.

In Chapter 4 Beers and Veldkamp argue that images of sustainability play a crucial part in innovating hard-, soft- and orgware. The way in which a 'dream' or a 'nightmare' is communicated with the use of images and metaphors, has a very strong impact in the outcome of the innovation process. If 'nightmare' images are dominant, then there is a good chance that the change will not happen. If, on the other hand, the positive dream is very strongly visualized, then the use of precisely that positive image can play a crucial role in engaging other stakeholders in the process.

Based on these observations in Chapter 4 the deliberate use of images is advocated. Observing that images can have a strong positive or negative influence leads to the conclusion that influencing the change may be best served by indeed managing the way in which images are used in the transformation process.

A first way to handle this is making use of *collaborative visualization* within the context of a project to explore the implicit assumptions that are inherent to images. Using different modeling approaches the authors show that by clarifying the contents of different images in this collaborative effort, a better understanding of the dynamic process of the systems involved will result. This observation hints into the direction of intervening in an innovation process by elucidate the images held by different stakeholders. This tool can be used in the process monitoring and reflection strategy that must be in place when dealing with a wicked innovation challenge.

A second way is *actively managing images* that are used by parties outside the project to communicate their ideas about the initiative. In almost all cases that deal with more wicked problems, multiple stakeholders are involved in influencing the general opinion about the project by using images that will benefit or threaten the innovation at hand. Being aware of these influences can help to frame or re-frame the project in such a way that it will have a positive effect on the innovative potential. If necessary, image management will even lead to engaging more societal actors as stakeholders in the project to incorporate their ideas.

The analysis in Chapter 2 also points into the direction of possibilities to actively stimulate inventions. Three different strategies are proposed by the authors. Dependent on the problem to solve, future inventions can be stimulated by organizing directional, intersectional or open innovation processes, sometimes followed by a switch between them.

The directional innovation process stems directly from an explicit omission or gap in the existing world. Especially in more technical terms it is possible to articulate in rather precise terms what is lacking. This can lead to targeted searches and investigations for inventions that will deliver the goods. The example given in Chapter 2 is the use of cisgenesis to develop new plant varieties with traits that serve both profit and planet aspects. Other examples can be found in the directional development of new stable systems in intensive livestock farming that have a much better performance with respect to animal health and even comfort.

The intersectional innovation process uses the explicit combinations of unlikely allies to speed up the 'Eureka!' moment. By deliberately organizing or constructing an intersection between different fields of expertise a place is created where ideas from different cultures, disciplines and fields meet causing many different inventions. As an example the development (or invention) of organic farming is described. This type of farming resulted from a new mix from such different fields as ecology, nature, human health, philosophy, ethics and religion. The idea of an intersection is also visible behind the idea of developing a 'Food Valley' in The Netherlands. Stimulated by the success stories from Silicon Valley in the USA, a lot of effort is put in bringing together international food companies, research institutes, business and financial services and institutions around Wageningen University. The aim of this regional intersection is to stimulate new cross-cutting inventions that will give rise to new business development.

The open innovation process uses the interaction between large groups of users and producers to come up with answers to questions and challenges that are put out in the open. Traditionally this was the way in which inventions in the agricultural domain were stimulated. However, the focus has been primarily on matters of productivity rise and lowering production costs. The same method can also be used to stimulate new ideas on realizing a better eco-efficiency or improving on issues related to human and animal health. The development of care farming is given as an example of using open innovation to come up with new arrangements in health care that make use of existing knowledge and know-how of farmers. Together they have developed a completely new system of day care for people with a mental disability, youth with behavioral problems, drug addicts and psychiatrically patients

For future developments in agro innovation, this has the interesting consequence that research supporting innovation should involve applied and tacit knowledge to a larger extend. This implies a different role of science, an increased involvement of applied colleges and companies besides the knowledge input of the classical universities. Such projects should be treated as an investment, including participation of the financial world and entrepreneurship from the outset.

7.3.2. The Social Sciences lens

In Chapter 3 Mommaas and Eweg present the sociological notions of regimes and niches to gain a better understanding of the processes involved in transformational changes. Experimentation in a great variety of niches will uncover basic institutional challenges that modern agriculture is faced with. Examples here are the growing tension between agriculture as part of a globalizing food economy, and agriculture as a source of landscape qualities in an urbanizing environment. In normal day-to-day (policy) practice, these two aspects of agriculture are treated and taken care of in separate domains and discourses. However, a real change in the direction of a more sustainable development asks for alignment of these different discourses. By actively bringing together a broader field of knowledge generation and innovation seeking practices in these niche experiments, we will enable the exploration and exploitation of new more sustainable value propositions that take into consideration all the different discourses. By adding up all these experiences in the various niche experiments the evolutionary transition in existing agro production regimes will be enhanced.

The analyses that underpin the storyline in Chapter 3 reveal that currently two major institutional blockages are hindering the more general introduction and application of system innovations that will lead to new 3P businesses. These two blockages are:

1. Ongoing difficulties in bringing alternative value propositions to consumer markets.
2. Increasing tensions between agro innovation trajectories and the changing role of the rural landscape in urbanizing environments.

These two issues ask for different approaches. The first blockage can be dealt with if we are able to open up the consumer markets for agricultural products and services that take into account ecological and social values. In Chapter 5 this challenge is taken up by Van Trijp and Fischer. Next to the vast body of literature already available on this topic, a number of experiments have been executed to address the potentials of bringing sustainability issues to the consumer. The conclusions from this analysis are ambiguous. On the one hand it is clear that for many innovation processes effective value capturing critically relies on willingness to buy and willingness to pay. For that reason, future agro innovation should also capture retailers and end consumers in the multi-stakeholder approach at an early stage to ensure that these consumer aspects are taken up fully while working on the system innovation. On the other hand, autonomous actual consumer demand for sustainable products and services is currently too weak to drive sustainable development at the speed that is needed on a commercial basis only. Therefore Van Trijp and Fischer suggest in their contribution that market creation for sustainable development needs to be part of agro innovation itself. They suggest that there has to be enough volume (by 'enlarging the pie') of supply and demand before the market can effectively self-regulate sustainable development. Enlarging this volume then becomes a major challenge for a collective multi-stakeholder effort within the trajectory of the system innovation. So, there is a potential here, but the innovation effort should be focused clearly on the market and marketing aspects.

The second blockage is even more difficult to address. What is needed here is a re-arrangement of the relationship between an agricultural and its physical environment. Following the increasing diversification of the demand – in quality and quantity of products, but also the environmental and social performance – the agricultural production system is also diversifying very rapidly. This puts a strain on the current ideas about organizing the spatial layout and integration of the constituting production entities. Old paradigms related to agricultural landscapes and spatial qualities have to be revised. The old ideal of pastoral landscapes that come with traditional modes of agricultural production must make way for a far more complex mosaic of production types and accompanying landscapes and spatial patterns. This will need an innovation of perception, judgment and appreciation of spatial qualities by large social groups. Mommaas and Eweg propose again a multi-stakeholder approach to take up this challenge.

7.4. A promising perspective

So, did we come up with a definite answer to the question that drives this research effort? Did we indeed identify what is needed to shift from a linear innovation approach to a new agro innovation system? As always, the answer to this question will be found somewhere in the middle. Yes, we did come up with a way to deal with the wicked problem of sustainable development, and yes we were able to translate our experiences into a set of principles and strategies that together make up the TransForum model. But we must also acknowledge that this is only a first effort in describing such a complex system as the non-linear and adaptive agro innovation system. Moreover, it may well be that other efforts will lead to partly comparable and partly differing approaches that seem to work in their context. So we have come up with an answer to the question, but we fully understand that differing answers are also possible.

In the TransForum program we were able to show that within the agricultural sector there is a true ability to cooperate in innovation processes. In the 30 action experiments cooperation with unlikely allies occurred. This cooperation led to interesting results based on mutual understanding and co-creation of solutions that appealed in some sense to all participants. In all experiments scientists, entrepreneurs, government officials and people from societal group worked together and came up with results that never would have been developed by either of these parties individually. To achieve that result it was necessary to actively work on engagement of the different stakeholders. The TransForum model with six principles and two core strategies proved very helpful in keeping the engagement alive and operational over the time span of the innovation experiments.

Another interesting observation was the need for concrete action perspectives for the different stakeholders in each of the action experiments. That did not mean that an overarching theme or challenge had to be developed. By accepting the different values and the accompanying visions and images that every participant took to the table, room was created to find new solutions that connected to the various value sets and orientations. This proved possible by effectively using the function of process monitor to keep track of the various value orientations and regularly calibrate the basis for cooperation.

This calls for a substantial effort in organizing and guiding the innovation effort. This effort must focus on the input that each of the participants must have in the innovation effort. With respect to the scientific community this calls for vigilance. Also in our experiments and accompanying research program we were not able to fully engage science and the scientific community in the innovation efforts. In some of the action experiments scientist did play a different and equal role in the execution of the experiment. However, with respect to the accompanying scientific research we were able to fill the research agenda with topics from the action experiments. The research itself, however, was carried out in a more traditional sense. It apparently is difficult to step outside the comfort zone of disciplines, even if the will and inclination is there to participate in engaged sustainability scholarship. In the research that was executed within the TransForum program the main driver was still the appreciation of the work by the peer review community.

For the future this constitutes a challenge. With respect to the engagement of the research community a lot more can be done and indeed needs to be done. We are only beginning to understand the intrinsic delicacies of innovation processes in situations where values, beliefs and stakes are widely differing. The shift from the linear innovation model also has repercussions for the role and responsibility that science plays in the innovation efforts. This role will become more complex and more diverse. Our experiences in the action experiments reveal that shifting this role can be very rewarding and may lead to extraordinary results for sustainable development.

7.5 References

Norton, B. G. (2005). *Sustainability: A philosophy of adaptive ecosystem management*. Chicago: University of Chicago Press.

Van Bueren, E.M., Klijn, E.H. and Koppenjan, J.F.M. (2003). Dealing with wicked problems in networks: Analyzing an environmental debate from a network perspective. *Journal of Public Administration Research and Theory*, 13(2), 193–213.

List of Abbreviations

3P	Triple P (people, planet, profit)
CGIAR	Consultative Group for International Agricultural Research
CSR	Corporate Social Responsibility
FAO	United Nations Food and Agriculture Organisation
GM	Genetic Modification
GMO	Genetically Modified Organisms
ILA	Interactive Learning in Action
IPR	Intellectual Property Rights
IRRI	International Rice Research Institute
KENGi	Knowledge institutions, Entrepreneurs and business partners, Non-governmental or civil society organizations, Government bodies, working together on innovation.
MOA	Motivation, Opportunity, Ability
NFW	Northern Frisian Woods
NGO	Non Governmental Organization
NMF	New Mixed Farm
PBR	Plant Breeders' Rights
PR	Patent Rights
RMA	Reflexive Monitoring in Action
TDNA	Transfer-DNA
TPB	Theory of Planned Behavior
UNEP	United Nations Environmental Program
UPOV	International Union for the Protection of New Varieties of Plants
WTP	Willingness to Pay

List of Contributors

Karin Andeweg Andeweg@TransForum.nl; TransForum, P.O. Box 80, 2700 AB, Zoetermeer, The Netherlands

Pieter J. Beers PJ.Beers@wur.nl; TransForum and Education and Competence Studies, Wageningen University, The Netherlands

Rik Eweg Eweg@TransForum.nl; TransForum, P.O. Box 80, 2700 AB Zoetermeer, The Netherlands

Arnout R. H. Fischer Arnout.Fischer@wur.nl; TransForum and Marketing and Consumer Behaviour, Wageningen University, Wageningen, The Netherlands

Evert Jacobsen Evert.Jacobsen@wur.nl; TransForum and Plant Breeding, Wageningen University, Wageningen, The Netherlands

Sander E. Mager Mager@TransForum.nl; TransForum, P.O. Box 80, 2700 AB, Zoetermeer, The Netherlands

Hans Mommaas J.t.mommaas@uvt.nl; TransForum and Telos, University of Tilburg, P.O. Box 90153, 5000 TE Tilburg, The Netherlands

H. Christopher Peterson Peters17@msu.edu; Nowlin Chair of Consumer-Responsive Agriculture, Michigan State University, USA

Tom Veldkamp Veldkamp@itc.nl; Faculty ITC, University of Twente, P.O. Box 6, 75AA Enschede, The Netherlands

Henk C. van Latesteijn VanLatesteijn@TransForum.nl; TransForum, P.O. Box 80, 2700 AB, Zoetermeer, The Netherlands

Hans C. M. van Trijp Hans.vanTrijp@wur.nl; TransForum and Wageningen University – Marketing and Consumer Behaviour Group; Hollandseweg 1, 6706 KN Wageningen, The Netherlands

Appendix I. TransForum: Stimulating innovation for sustainable development by learning by doing

Let's stop talking about sustainable development, start working on it. With that in mind, TransForum started as a 6-year innovation program to stimulate the transition toward a more sustainable development of the Dutch agricultural sector. Since there is not one definite formulation of sustainability, there neither is there one true solution. Rather than debating on the definition and attempting to seek ultimate sustainability, TransForum tackled this issue by defining sustainable development as an ongoing process of improvement. We started experimenting with different and new forms of agricultural activity that created a more sustainable situation, meaning that the experiment should, in comparison with the current situation, improve on issues related to economic performance (profit), impact on the environment (planet) and compliance with social values (people). We worked together with people from the industry, from local governments, from societal organizations and from knowledge institutes in order to develop concrete business cases in which the next step in sustainable development was envisaged.

In 6 years TransForum conducted a various set of over 30 action experiments, complementary scientific research and learning projects. Figure I shows the structure of the program and coherence between the projects.

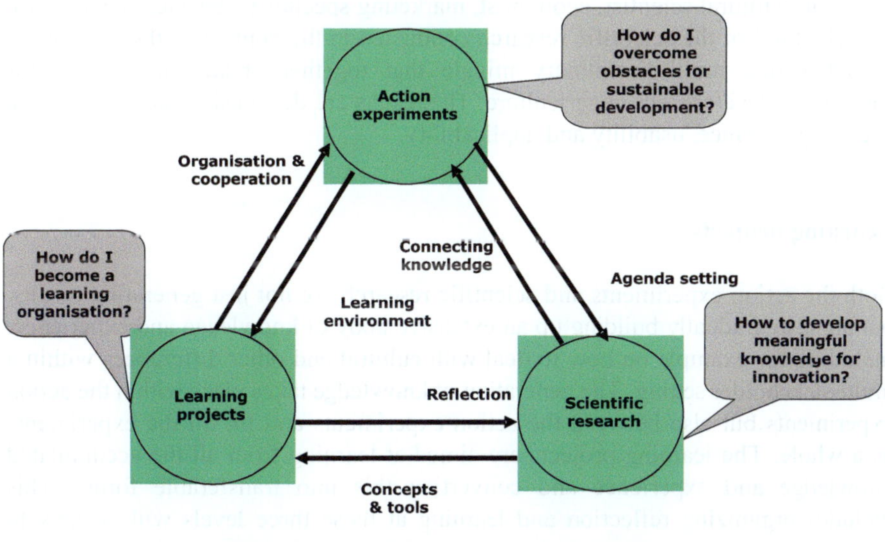

Fig. I. Structure of the TransForum program.

Action experiments

The current agricultural debate is often dominated by two paradigms, organic vs. industralised, small-scale vs. large scale agriculture, etc. However, according to many (e.g. Foley 2009)[1], neither can fulfil the global needs with respect to a sufficient food supply and in the mean time also meet the sustainability goals. Approaches with aspects from both sides are necessary. The action experiments therefore cover diverse types of new and more sustainable agriculture. They vary from experimenting with saline agriculture to creating so-called 'agroparks'. From combining new forms of care with agriculture to creating new (international) food supply chains, to regional oriented experiments to combine agriculture and landscape management. In all action experiments, different stakeholders worked together to realize a more sustainable development of their agricultural production. Appendix II gives more information of the action experiments that have been addressed in the chapters of this book.

Scientific research

Out of the action experiments, persistent issues appeared on how innovation for sustainable development can be stimulated. Scientific research complemented the action experiments with new knowledge, solutions and instruments to enhance innovation. Most of the scientific research has been conducted in the context of the action experiment. This required scientists from different disciplines: socio-logist, agricultural scientist, economist, marketing specialists, business economics, etc. The total of the scientific research conducted in the context of the program is an interesting multi-disciplinary mingle that together created new ideas for sustainable development in agriculture. The results are designed to have immediate social significance, usability and applicability.

Learning projects

Both the action experiments and scientific research are not just generating results, but are also gradually building up an extensive body of knowledge and experience, including for example on how to deal with cultural and other differences within a multi-stakeholder setting. The generation of knowledge takes place within the action experiments but also between the action experiments and for all the experiments as a whole. The learning projects are aimed at learning from all the accumulated knowledge and experience and converting this into transferable forms. This includes organizing reflection and learning at these three levels with a view to

---

[1] Foley, J. (2009). The Third Way. In *Can Biotech Food Cure World Hunger?*, New York Times, Room for Debate Blog. NYTimes.com.

identifying the competencies and skills needed for the realization of innovations and a transition.

For more information an extended elaboration on TransForum's approach, see also: Veldkamp et al. 2008.[2] An overview of all the experiments and projects can be found in Appendix II.

[2] Veldkamp, A., Altvorst, van A.C., Eweg, R., Jacobsen, E., Kleef, van A., Latesteijn, van H.C., Mager, S.E., Mommaas, H., Smeets, P.J.A.M., Spaans, L., Trijp, van J.C.M. (2008). Triggering transitions towards sustainable development of the Dutch agricultural sector: TransForum's approach. *Agronomy for Sustainable Development*, 28, 87–96.

Appendix II. Action experiments explicated

The research presented in this book is conducted in context of the action experiments of the TransForum program. Eight action experiments were addressed specifically throughout the book. These are presented in more detail below. A short overview of all the action experiments can be found in Appendix III. More information can be found at the TransForum website: www.transforum.nl.

Action experiment: SynErgy

The action experiment SynErgy experimented together with growers, researchers, and governments with the energy producing greenhouse, a closed green house that can produce energy rather than only utilizing energy. In such a closed greenhouse, growing condition differ. During the action experiment, bottlenecks concerning plant physiology had to overcome. An integral crop growth model was needed that includes water status and its effects on physiological processes. This required self-learning models that used sensor information for online updating model results. To realize this, a community of practice was developed, in which the growers shared experiences with and posed questions to researchers (Hoes et al. 2008).[1] Cooperation and active exchange of knowledge proved to be necessary to help all participants of the platform to increase understanding and to apply the information and realize their goals.

Action experiment: More about Food

In order to bring about more sustainable development in the agro sector, consumer demand for sustainably produced food must increase. The majority of consumers, however, are not really aware of how their food is produced. This action experiment tests whether providing better information about the sustainability performance of our food increases the knowledge and consumption of more sustainably produced products.

Consumers can find information on the website about food production methods and can compare the sustainability performance of the respective methods. Participating producers and retailers can obtain valuable information through the website about consumer needs, and use this information to develop new and more sustainable food concepts. Consumers are offered an instrument to learn more about sustainably produced food and to adjust their purchasing behavior on the basis of this information. The experiment also links consumer profiles to more sustainable purchasing behavior so as to facilitate targeted marketing.

---

[1] Hoes, A.C., Regeer, B.J. and Bunders, J.F.G. (2008). Transformers in knowledge production: building science-practice collaborations. *Action Learning: Research and Practice, 5,* 207–220.

Action experiment: Regional Food Chain

Farmers in the vicinity of Amsterdam have joined forces to establish a regional food chain for the city of Amsterdam. Collectively they are offering their products directly to the consumer via alternative distribution channels rather than via a retailer. They sell their fresh products in a separate corner of a shop in a popular shopping area. In this business model, the producer remains the owner of the product even after it reaches the supermarket shelf. Thereby they establish direct contact between producer and consumer. Consumers can consult a website to see where the products come from and how they are produced. They can even visit the farm. In this way, consumers are more aware of the food and the way it is produced. And the farmer receives a fair price for high-quality products which are sustainably produced.

Action experiment: Green Care

Dealing with nature and the day-to-day chores and work on a farm can be deeply healing for people suffering for example from burnouts, psychosocial complaints, labor market reintegration, or children with behavioral problems. This is especially suitable for people who live in urban areas. For farmers close to urbanized areas it can be an attractive proposition to develop side activities in addition to their daily farming activities. This experiment had brought together farmers, health care insurers, health institutes government authorities and researchers to develop professional and profitable range of care facilities around the city of Amsterdam (Hassink et al. 2007).[2] Clients were interested in labor activities on farms, the insurance companies and welfare organizations regard care farming as a less expensive but more effective form of rehabilitation, which fits in the financial health care system (aimed at cost reduction). In this way, farmers have additional income (profit), needs of urban dwellers are served (people) and the countryside around cities can better resist the growing urban expansion preserving green areas in urbanized areas (planet).

www.landzijde.nl www.groenehoed.nl

Action experiment: New Markets and Vital Coalitions Heuvelland

Where agriculture is disappearing, the loss of the related landscape is a threat. The aim of this experiment is to secure and strengthen the environmental qualities of

---

[2] Hassink, J., Zwartbol, Ch., Agricola, H., Elings, M., and Thissen, J.T.N.M. (2007). Current status and potential of care farms in the Netherlands. *Netherlands Journal for Agrarian Sociology, 55*(1), 21–36.

the landscape by achieving a new investment coalition of economic carriers able to sustainably support the characteristic countryside (Van der Stoep and Aarts 2010).[3] New economic activities such as recreation, hotels, restaurants and catering facilities, care and health can ensure that the typical hilly landscape in the south of the Netherlands is retained. Using these opportunities is essential for the further economic development of the province of Limburg.

These opportunities can only be used if players in the area cooperate with one another. In this action experiment new product-market combinations are developed jointly with entrepreneurs, the authorities and knowledge institutions that will support the landscape. The innovative side of the project is that links are being made with new players, such as banks, project developers, healthcare insurers and hospitals. The action experiment searches for organizational forms that will really allow 'red' to take responsibility for 'green'.

Action experiment: New Mixed Farm

In the New Mixed Farm farmers go back to the old integrated farm of the beginning of the former century, but in a modern way. Entrepreneurs cooperate together to realize a more sustainable business in terms of economically viable, environmental efficient and improve animal welfare.

The New Mixed Farm constitutes a closed poultry farm ranging from egg right through to chicken fillet, a closed pig farm from parent sow through to meat pig and an installation firm to generate heat and energy. The NMF is an example of an agro-park. An agro-park is an innovative form of industrial farming that uses high-tech innovations to reduce costs and environmental damage by combining a number of distinct agricultural businesses to enable the reuse of resources such as energy, gas, minerals, and manure. In the case of the NMF this will lead to 60–80% less energy use, 30–40% reduction of greenhouse gas emissions, decrease in animal transport and a higher profitability.

Action experiment: Saline Agriculture

Pressure is growing on the ever more limited supplies of fresh water. Also in the Netherlands saline seepage is on the increase with the rise of sea level and the sinking of the old polders due to climate change and subsidence. Saline agriculture experiments with salt-tolerant crops in the coastal area of one of the islands (Texel). Researchers and growers are collaborating to generate new knowledge for sustainable, innovative and profitable crops and cultivation methods. This is an

---

[3] Stoep, H. van der and Aarts, M.N.C. (2010). New markets in Heuvelland: coalition building and agenda setting. In: I. Horlings (Ed.), *Vital coalitions, vital regions. Partnerships for sustainable, regional development*, Chapter 6 (pp. 121–150). Wageningen: Wageningen Academic Publishers.

example of how farmers can adapt to changing circumstances like climate change instead of combating such change, and thereby reducing environmental impact by using few if any pesticides. In the mean time this new way of growing is promoted in order to create market demand. Sea kale, barley, spelt, quinoa and rocket have been launched in the market and in quality restaurants.

www.ziltezeekool.nl

Action experiment: Northern Frisian Woods

The Northern Frisian Woods, in the northern part of The Netherlands, is an exclusive national landscape of 55,000 hectares with a combination of historic-cultural elements and natural elements. The farmers in the area are important managers for the maintenance of this unique small-scale landscape. Preservation of the area used to depend on the aid of government subsidies. Now, researchers, local, regional and national governments, societal organizations and entrepreneurs are working together on the management and development of the area to create an improvement in the profitability of the local farms while also establishing greater sustainability and a continued preservation of the landscape.

One of the challenges in preserving the small-scale landscape is to comply with existing national and European manure regulation. Moreover, they started experimenting with manure application by spreading it on the land, instead of injecting it as prescribed by Dutch legislation. The stakeholders together created new knowledge for the development of a 'closed loop farming' system. The farmers aimed for a reduction of nitrogen in the manure through using less protein and fiber in feed and fodder (Stuiver 2008).[4] By means of a modified diet for the cows, locally produced feed and the use of manure on the farmers' own land, a system of more sustainable farming will be developed that is consistent with the small-scale landscape while also producing even higher quality milk.

www.noordelijkefriesewouden.nu

---

[4] Stuiver, M. (2008). Regime change and storylines: A sociological analysis of manure practices in contemporary Dutch dairy farming. Dissertation. Wageningen: Wageningen UR.

Appendix III. Overview experiments and projects

Below you find an overview of all action experiments, scientific research projects and learning projects that have been conducted within the TransForum program.

Action experiments

All action experiments listed below have been conducted within the TransForum program. Within the experiments knowledge institutes, entrepreneurs, societal organizations and governments worked together to create new knowledge for innovation and sustainable development. More information on the all the projects can be found at www.transforum.nl.

1. Accessibility of Mainport Aalsmeer*
2. Agropark Flevoland
3. Biopark Ghent-Terneuzen
4. More about Food
5. Calendula
6. C2C and Logistics
7. Dairy Adventure
8. Drive*
9. Flor-i-Log
10. Freshpark VENLog
11. Green and the City
12. Green Care Amsterdam
13. Greenport Betuwse Bloem
14. Greenport Shanghai
15. Green Valley
16. Healthy Hard-Fruit Chain
17. Healthy Oats
18. International Livestock Farming Orchestration Role*
19. LandMarkt
20. Mainport – Greenport
21. New Markets and Vital Coalitions Heuvelland
22. New Mixed Farm
23. Northern Frisian Woods
24. Regional Food Chain
25. Saline Agriculture
26. Saline Experimental Garden
27. Special Education on the Farm
28. Streamlining Greenport Venlo
29. Sustainability in Retailing

30. Synergy
31. The Quest for the Golden Egg
32. The Sjalon
33. Water Self-sufficient Agriculture

*Action experiment did not succeed and stopped early on in the process.

Scientific research projects

1. Knowledge management in International Networks
2. The first database
3. 3MG
4. Historical and future transitions in Agriculture and Food
5. Stacking functionally expressed apple genes
6. Images of sustainable development of Dutch agriculture and green space
7. Understanding company level marketing strategies for sustainability
8. Stimulating consumer choice for sustainable products at the points of purchase
9. Vital Coalitions
10. Effects and effectiveness of agriculture and healthcare
11. Robustness of animal production systems
12. SynErgie: Crop of the future in Greenhouse of the future
13. Self-organization of innovation networks
14. Sustainability and dynamics of food consumption patterns
15. Drivers for successful transitions to new activities by farmers
16. Identifying critical stakeholders for sustainable products
17. Conceptualization of robust production systems
18. Networking and governance strategies in agricultural innovation networks
19. Improving interaction between KENGi partners
20. Subtle activation of sustainability values
21. Toolbox for value creation in innovation networks
22. SUSMETRO
23. Economic risk analysis of agroparks
24. Green Care business models
25. Inner angels and demons

Learning projects

1. All about Food
2. Business cases for transition projects
3. Cyclical Innovation Model (CIM)
4. Education – Analysis & Problem Exploration
5. Evaluation of Innovative Entrepreneurship Learning Process
6. Film-plan for Sustainable Agriculture Images

7. International Association for Sustainable Agroparks
8. Mode-2 Publication Strategy
9. Netchain Game
10. Network managers 'The Art of Cooperation'
11. Open innovation in Agriculture
12. People Management
13. Kwartiermaken PIAMA
14. Project Direction Development
15. Regional Development Community of Practice
16. Steering knowledge productivity
17. Swaminathan Foundation Seminar
18. Task Force Metropolitan Agriculture
19. TransForum as Mode-2 Intermediary
20. Agroparks International process monitoring
21. Agropark Lepakshi Knowledge & Competencies
22. Business Cases TransForum
23. Connected Value Development
24. Embedding of 3P incentives
25. Greenport Shanghai process evaluation
26. MetroAg Innoversity Summit
27. Multi-Actor Process Management
28. Networked Learning and Learning From Networks
29. Re-imagining Urban Systems: Towards a Vision of Metropolitan Agriculture
30. Action experiments sustainability scan
31. TransForum Business Cases
32. Transition Monitoring Pizza project
33. TransLearning
34. ZLTO 'Role of Collectives' Innovation System

Appendix IV

Attributes used in the analysis of the 30 action experiments (see Chapter 6)

Internal Context	External Context	Process
Initiating Conditions 1. How the project was initiated 2. Who initiated the project 3. Why the project was initiated *Project Membership* 4. Who engaged in the project 5. Who was not engaged in the project 6. How "who" changed across project *Wickedness* 7. Level of complexity/ uncertainty of problem 8. How well defined the problem was 9. How well understood cause-and-effect relationships were *Potential for Conflict* 10. The number of issues dividing project participants 11. The number of issues uniting project participants 12. Level of agreement among project participants 13. Level of disagreement among project participants	*External Conditions* 1. Obstacles related to economic conditions 2. Opportunities related to economic conditions 3. How economic conditions changed across the project's life 4. Obstacles related to political/regulatory conditions 5. Opportunities related to political/regulatory conditions 6. How political/regulatory conditions changed across project's life 7. Obstacles related to social and community conditions 8. Opportunities related to social and community conditions 9. How social & community conditions changed across project's life 10. Obstacles related to location and physical setting 11. Opportunities related to location and physical setting. 12. How location and physical setting changed across project's life	*Initiating Conditions* 1. How intentionally the process was selected 2. How organically the process arose as project unfolded 3. How the process was initially managed *Leadership* 4. Who led the process and how 5. How process leadership evolved over time *General Process Management* 6. How process management evolved over time 7. How the 3P impacts were considered 8. How the vision and objectives were formulated 9. The strategies used to manage obstacles and unexpected events 10. Strategies used to communicate with critical actors inside project 11. The strategies used to communicate with critical external actors 12. The strategies used to align the project with its external context *Innovation Management* 13. Strategies used to reconcile differing incentives across team 14. "Translation" work done to deliberately find common language

14. Inherent level of value conflict among the participants 15. Inherent level of incentive conflict among the participants 16. The level of trust among participants *General Project Traits* 17. The vision for the project 18. The project's objectives 19. Riskiness of the project's outcomes or implementation 20. Effectiveness of internal project champions 21. Inherent strengths of the project 22. Inherent weaknesses of the project *Size and Scope* 23. The project's scale (size of what had to be done) 24. The project's scope (breadth of what that had to be done) 25. The project's divisibility into smaller more manageable parts 26. The project's divisibility into smaller less risky parts 27. The project's divisibility into activities across time	13. Obstacles related to environmental conditions 14. Opportunities related to environmental condition 15. How environmental conditions changed across the project's life *External Actors* 16. The influence of specific external actors opposed to the project 17. The influence of specific external actors in favor of the project 18. The absence of influence by specific external actors 19. The project's riskiness for external actors 20. The influence of external project champion(s) 21. Extent of external demands, expectations or requirements 22. Media attention given to the project	15. "Translation" work done to understand/meet various incentive needs 16. How existing explicit knowledge was used 17. How existing tacit knowledge was used 18. How experimentation in action occurred 19. How process phases were used 20. The strategies used to create innovation 21. The strategies used to create action learning 22. The strategies used to generate new knowledge 23. The strategies used to create transdisciplinary collaboration *Impact of Intermediary* 24. Function(s) assigned to intermediary, including TransForum 25. How the intermediary carried out its (their) function 26. How the intermediary function(s) changed across time 27. When in the life of TransForum the project occurred

MIX
Papier aus verantwortungsvollen Quellen
Paper from responsible sources
FSC® C105338

If you have any concerns about our products,
you can contact us on
ProductSafety@springernature.com

In case Publisher is established outside the EU,
the EU authorized representative is:
Springer Nature Customer Service Center GmbH
Europaplatz 3, 69115 Heidelberg, Germany

Printed by Libri Plureos GmbH
in Hamburg, Germany